Silicon-Based Structural Ceramics for the New Millennium

Related titles published by The American Ceramic Society:

For information on ordering titles published by The American Ceramic Society,
or to request a publications catalog, please contact our Customer Service
Department at:

Customer Service Department
735 Ceramic Place
Westerville, OH 43081, USA

614-794-5890 (phone)
614-794-5892 (fax)
customersrvc@acers.org (e-mail)

Visit our on-line book catalog at www.ceramics.org.

Ceramic Transactions
Volume 142

Silicon-Based Structural Ceramics for the New Millennium

Proceedings of the Silicon-Based Structural Ceramics for the
New Millennium symposium, held at the 104th Annual
Meeting of The American Ceramic Society April 28–May 1,
2002, in St. Louis, Missouri.

Edited by
Manuel E. Brito
National Institute of Advanced Industrial Science and
Technology

Hua-Tay Lin
Oak Ridge National Laboratory

Kevin Plucknett
Advanced Ceramics Group, QinetiQ

Published by
The American Ceramic Society
735 Ceramic Place
Westerville, Ohio 43081
www.ceramics.org

Proceedings of the Silicon-Based Structural Ceramics for the New Millennium symposium, held at the 104th Annual Meeting of The American Ceramic Society April 28–May 1, 2002, in St. Louis, Missouri.

Cover photo: "Fracture surface of exposed Kyocera SN235 sample after flexural testing at 850°C and 0.003 MPa/s" is courtesy of H.T. Lin, T.P. Kirkland, and A.A. Wereszczak, and appears as figure 6(a) in their paper "Effect of Long-term Immersion Test on Mechanical Reliability of Candidate Silicon Nitride Ceramics for Diesel Engine Applications," which begins on page 261.

For information on ordering titles published by The American Ceramic Society, or to request a publications catalog, please call 614-794-5890.

4 3 2 1–06 05 04 03
ISSN 1042-1122
ISBN 1-57498-157-9

ntents

Improved Properties

Applications

Preface

The international symposium, Silicon-Based Structural Ceramics for the New Millennium: Recent Scientific and Technological Developments, was held during the 104th Annual Meeting of The American Ceramic Society (ACerS), April 28–May 1, 2002 in St. Louis, Missouri. This symposium focused on recent scientific and technological developments in silicon-based (i.e., silicon nitride, SiAlONs, silicon carbide, silicon oxynitride) structural ceramics. Attendees from academia and industry assessed the current state of the art and industrial case studies were advocated to highlight the development and application of these materials in real engineering environments. The symposium consisted of 55 contributions from 12 countries and five continents.

This volume of the Ceramic Transactions series contains selected papers on various aspects of synthesis and processing, characterization, properties and applications of silicon-based structural ceramics. The manuscripts were reviewed following The American Ceramic Society review process.

The editors wish to extend their gratitude to all the authors for their contributions and to all the participants and session chairs for their time and efforts. Thanks to all the reviewers for their useful comments and suggestions. Thanks are due to the staff of the Meetings and Publications Departments of ACerS for their tireless efforts. We especially appreciate the helpful assistance, cooperation and patience of Mary Cassells and Greg Geiger throughout the production process of this volume.

It is our hope that this volume will serve as a useful reference for scientist and engineers interested in the field of structural ceramics, and also will serve as a stimulus for continuing efforts in the implementation and wider use of silicon-based ceramics.

Manuel E. Brito

Hua-Tay Lin

Kevin Plucknett

Novel Synthesis and Processing

COLLOIDAL PROCESSING OF SILICON NITRIDE

Eric Laarz, Anders Meurk and Lennart Bergström
Institute for Surface Chemistry
PO Box 5607
SE-11486 Stockholm
Sweden

ABSTRACT

This review summarizes recent work related to the colloidal processing of silicon nitride. Silicon nitride surface chemistry, dissolution behavior, deagglomeration, dispersion and stabilization in aqueous medium are treated in detail, taking into account apparent interdependencies. Experimental results have been evaluated quantitatively and qualitatively to identify underlying mechanisms responsible for colloidal properties of silicon nitride powders. Fitting dissolution data at various pH and temperatures with suitable kinetic equations enabled us to determine the overall activation energy for dissolution and to identify different kinetic dissolution regimes. Evidently, the dissolution behavior of silicon nitride surface oxides and amorphous silica are very similar. It is further shown how milling conditions for the deagglomeration of silicon nitride powder containing hard particle agglomerates with inter-particle necks can be optimized. Forces between silicon nitride surfaces were measured with an atomic force microscope and evaluated using Lifshitz and DLVO theory and numerical data analysis. This approach allowed the verification of Hamaker constants for van der Waals interaction across various media as well as an assessment of charge regulation effects at varying pH and surface separation. By measuring forces between Si_3N_4 surfaces with adsorbed polyacrylic acid dispersant, polymer-induced steric forces were shown to be very short-ranged. Hence, the dispersant's effect on colloidal stability is limited to a purely electrostatic contribution to the total particle interaction potential.

CONTENTS

INTRODUCTION

Present methods of manufacturing ceramic green bodies usually start with a suspension where the ceramic particles (powders, whiskers, platelets, etc.) are mixed with a liquid or a polymer melt, a proper dispersant, and possibly further additives (such as binders, plasticizers, and anti-foaming agents). Colloid science is frequently employed in order to remove heterogeneities and to optimize the suspension rheology [1, 2], e.g. by manipulation and control of the interparticle forces in powder suspensions [3, 4]. Hence, colloidal processing has become an integral part in the development and optimization of industrial processes for the production of technical ceramics.

This review describes some fundamental aspects of the colloid and surface chemistry of silicon nitride with importance for colloidal processing of aqueous suspensions. To support the understanding of the underlying colloidal concepts, we briefly introduce a background on the surface chemistry and powder dissolution. This is followed by a section on deagglomeration and dispersion. An orientation how the rheological properties of concentrated suspensions can be manipulated is also given.

SURFACE CHEMISTRY AND POWDER DISSOLUTION

The dissolution behavior of ceramic powders can play a crucial role in colloidal processing for several reasons. Changes of the elemental surface composition in the course of dissolution or dissolution/precipitation processes strongly affect the particle surface chemistry, which will modify interparticle forces and adsorption behavior of organic additives. Similarly, the release of ionic species due to dissolution increases the ionic strength in solution, which also alters the electrostatic interaction potential and adsorption processes. As further discussed in the following chapter, dissolution can even have a pronounced influence on agglomeration and deagglomeration processes.

Probably all ceramic materials used in technical applications dissolve to some extent when exposed to an aqueous environment. The dissolution rate, however, can vary by several orders of magnitude depending on the chemical composition and structure of the solid and the solution conditions. Hence, some ceramics are readily dissolved, while others exhibit a negligible dissolution rate. However, the dissolution rate is usually strongly temperature and pH dependent. Due to the large number of parameters that influence the dissolution, with many of them not directly accessible by experimental methods, there is a lack of comprehensive theories that can predict the dissolution behavior of ceramics in water. Consequently, it is common practice to describe the dissolution of a specific material in terms of experimentally determined dissolution rate constants.

In our discussion we place focus on the surface chemistry and dissolution kinetics of silicon nitride powder at alkaline pH. It will be shown that there is a strong interdependency of dissolution and surface chemistry, with a pronounced effect on colloidal stability and deagglomeration. After giving a brief account of earlier investigations on the surface chemistry of silicon nitride, thermodynamic and kinetic aspects of hydrolysis and dissolution in aqueous suspensions are addressed in detail. Experimental data for the dissolution kinetics at different temperatures and pH are presented and the overall activation energy for dissolution is determined. Additionally, similarities in the dissolution behavior of silicon nitride and silica are discussed.

Surface Chemistry

The surface chemistry of silicon nitride is well described by the dissociation and protonation reactions of amphoteric silanol (Si-OH) and basic secondary amine (Si$_2$-NH) surface groups in water [5]

$$SiOH \xleftrightarrow{K_{a1}} SiO^- + H^+ \tag{1 a}$$

$$SiOH_2^+ \xleftrightarrow{K_{a2}} SiOH + H^+ \tag{1 b}$$

$$Si_2NH_2^+ \xleftrightarrow{K_{a}} Si_2NH + H^+ \tag{1 c}$$

where K_{A1}, K_{A2} and K_B are equilibrium constants. The surface reactions and the surface densities of the two surface groups determine the pH-dependent surface charge density of silicon nitride and, thus, the isoelectric point, pH_{iep} (i.e. the pH where the net surface charge is zero).

Based on Equation 1, Bergström and Bosted [5] derived an expression that relates the pH_{iep} to the relative amount of silanol, N_A, and secondary amine groups, N_B,

$$\frac{N_B}{N_A} = \frac{-\left(\dfrac{10^{-pH_{iep}}/K_{A1} - K_{A2}/10^{-pH_{iep}}}{1 + 10^{-pH_{iep}}/K_{A1} + K_{A2}/10^{-pH_{iep}}} \right)}{\left(\dfrac{10^{-pH_{iep}}/K_B}{1 + 10^{-pH_{iep}}/K_B} \right)} \qquad (2)$$

Relatively good agreement between N_B/N_A ratios calculated with Equation 2 and N/O-ratios obtained by XPS-measurements on powder surfaces was reported [5].

Earlier investigations have already demonstrated that the pH_{iep} of Si_3N_4 powders can vary dramatically depending on the powder synthesis process, powder treatment (thermal, chemical or mechanical), and wet processing conditions [5-15]. This has been attributed to differences in the oxygen content of the powder surface, which is mainly dictated by oxidation in air and hydrolysis reactions in water of the thermodynamically unstable Si_3N_4.

The oxidation of silicon nitride in air proceeds according to

$$Si_3N_4 + 3O_2 \xleftrightarrow{K} 3SiO_2 + 2N_2 \qquad (3)$$

when silica is assumed to be the main oxidation product. Investigations on pure Si_3N_4–films confirmed the presence of an amorphous, oxygen-rich surface layer with the oxygen concentration gradually increasing towards the surface [16]. The hydrolysis of silicon nitride in aqueous environments can be described with the overall reaction

$$Si_3N_4 + 6H_2O \xleftrightarrow{K} 3SiO_2 + 4NH_3 \qquad (4)$$

assuming that silica is the main hydrolysis product. Similar reactions can be written when silicon oxynitride is the oxidation product. The silicon oxide

Silicon-Based Structural Ceramics

reaction product is present in the form of an amorphous surface layer, whereas the ammonia dissolves in water and protonates according to

$$NH_4^+ \xleftarrow{\quad K \quad} NH_3 + H^+ \tag{5}$$

Dissolution Behavior

The oxidized surface layer on silicon nitride is soluble in water and leaching studies have shown that not only oxidation and hydrolysis but also dissolution may significantly alter the surface composition and the associated surface charge [5, 12, 13]. The effect of oxidation and dissolution on the surface chemistry of silicon nitride is reflected in electrokinetic zeta-potential measurements and XPS-measurements on oxidized and leached powders as shown in Figure 1 and Table I. The leaching treatment consisted of dialyzing a dilute powder suspension against 0.01 M NaOH solution. An oxidation treatment of the as-received powder (UBE SN-E10, Ube Industries, Japan) shifted the pH_{iep} to acidic values, whereas the leaching treatment caused a shift to basic values (Figure 1).

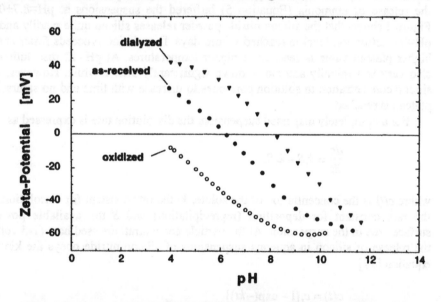

Figure 1: Zeta-potential of Si_3N_4 suspensions with 5 vol% solids loading: as-received powder (●), after oxidation in air for 8 hours at 800 °C (O), and after dialyzing against 0.01 M NaOH for several days (▼). The counterion concentration was $[Na^+]=0.01$ M in all experiments.

Both treatments lead to corresponding changes in elemental O/Si and N/Si ratios of the surface layer (Table I). These results agree with the suggested interdependency of the surface layer oxygen content, the relative amount of Si-OH and Si_2-NH surface groups, and the surface charge density.

Table I. Results of XPS-measurements on as-received and modified silicon nitride powders used for the zeta-potential measurements shown in Figure 1.

Sample	N/Si	O/Si	N (1s) binding energy	Si (2p) binding energy
as-received	1.03	0.59	394.4 eV	99.0 eV
oxidized	0.85	0.80	394.6 eV	99.2 eV
dialyzed	1.33	0.19	397.7 eV	101.7 eV

The kinetics of silicon nitride dissolution was evaluated by studying the release of silicon as a function of time at varying pH and temperature (Figure 2). Except for the experiment conducted at pH=12, where NaOH was added, all dissolution experiments were performed without pH-adjustment. In these cases the release of ammonia (Equation 5) buffered the suspensions at pH=8.7±0.2. Figure 2 shows that the silicon nitride powder releases silicon quite readily and at pH=9 a saturation level is reached within days. Dissolution proceeds faster and a higher plateau value is reached at higher temperatures. At pH=12 dissolution is also very fast initially and slows down significantly after a while. However, the silicon concentration in solution continues to increase with time and no saturation plateau is reached.

For a completely dispersed suspension, the dissolution rate is expressed as

$$\frac{dc}{dt} = k_1 S - k_2 S c \tag{6}$$

where $c(t)$ is the concentration of the solute, k_1 the rate constant for dissolution, k_2 the rate constant for deposition (reprecipitation) and S the available powder surface area in the suspension. At the particle concentrations used here (1-5 vol%), the release of silicon in aqueous suspensions of silicon nitride obeys the kinetic equation [17]

$$c(t) = c_s[1 - \exp(-kt)] \tag{7}$$

with c_s being the saturation concentration and k being the dissolution rate constant expressed in units of s^{-1}.

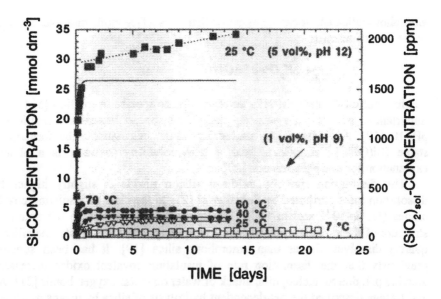

Figure 2: Kinetics of silicon release in dilute suspensions (1-5 vol%) of silicon nitride powder at pH=9 and pH=12. The experimental points are approximated by Equation 7 (solid lines). The dotted line is a least-square fit to the experimental data points. (from reference [18]]

The dependence of the dissolution rate constant, k, on temperature, T, can be described by an Arrhenius-type equation

$$k = k_0 \exp(-\frac{E^0}{RT})$$

(8)

where k_0 is the pre-exponential frequency factor, E^0 the activation energy, and R the gas constant. By employing Equation 8 we can analyze the silicon release at pH=9 at various temperatures (Figure 2) in order to determine the overall dissolution activation energy. A least-square fit to the $\ln k$ vs. $1/T$ data yields $k_0=9.8\times10^3$ s^{-1} and $E^0=51.7\pm0.1$ kJ mol^{-1}. Apparently, the obtained activation energy for dissolution of oxidized silicon nitride is close to values reported for amorphous silica (70-80 kJ mol^{-1}) [19]. Moreover, the experimentally determined amount of silicon dissolved from the oxidized silicon nitride surface at room temperature is around 170 ppm (Figure 2), which is also similar to reported values for amorphous silica, $[SiO_2]_{sol}\approx100-150$ ppm at pH=3-9 [19]. Hence, from a dissolution point of view, the oxidized surface layer of silicon nitride behaves like amorphous silica and it can be assumed that the surface oxide dissolves according to the overall reaction

$$SiO_2 + 2H_2O \leftrightarrow Si(OH)_4 \qquad (9)$$

yielding colloidal silica Si(OH)$_4$ or other silicate species in solution [5, 17]. It is important to note that the solubility of silicon species increases dramatically at pH>10 [19]. As indicated by molecular orbital calculations, the formation of stable $[Si(OH)^5]^{-1}$ complexes with a high solubility in water is a plausible explanation for this phenomenon [20].

The dissolution rate for oxidized silicon nitride is slightly higher than dissolution rates attributed by Segall *et al* [21] to the class of insulating covalent oxides (*i.e.* $k<10^{-11}$ mol m^{-2} s^{-1}). This is most likely related to the amorphous structure of the surface layer; Iler, for example, reported that crystalline silica (α-quartz) dissolves slower than amorphous silica [19]. It has been suggested previously that the dissolution rate of insulating covalent oxides increases at alkaline pH, due to nucleophilic attack of water on metal-oxygen bonds [21]. Xiao and Lasaga described the pH-dependent hydrolysis of silica by means of *ab initio* quantum mechanical simulations [22, 23]. Their calculations suggest that the hydroxyl ion acts as a catalyst for the hydrolysis of siloxane bonds; hydrolysis via nucleophilic attack of water on Si-O-Si bridges is affected by the protonation state of the silanol surface groups. The activation energies for hydrolysis of silica are 121, 100, and 79 kJ mol^{-1} for the cases where the respective surface silanols at the Si-O-Si bridge are neutral (*Si-OH*), protonated ($Si-OH_2^+$), or deprotonated (*Si-O⁻*). It was further suggested that hydroxyl-catalyzed hydrolysis involves the formation of a transitional penta-coordinated Si-complex prior to the subsequent cleavage of the bridging siloxane bond. In fact, the hydrolysis of silica is controlled by the formation of the transitional complex. This step requires the highest activation energy of 79 kJ mol^{-1}, the final rupture of the Si-O-Si bond involves an energy barrier of only 19 kJ mol^{-1} [22, 23].

The above information allows a more detailed discussion of the peculiarities of silicon nitride dissolution at pH=12 and T=25 °C. Figure 2 shows that data points of the initial dissolution stage at this pH can be fitted with Equation 7. In contrast to the dissolution at pH=9, no saturation plateau is reached at longer times. Instead, one observes a slow dissolution characterized by a constant increase in concentration of dissolved silicon. The absence of a saturation plateau can be explained by the formation of penta-coordinated Si-complexes in solution and the corresponding increase in silicon solubility. The estimated dissolution rate constants for the initial and the later stage dissolution at pH=12 differ by three orders of magnitude, with the latter one being much lower than the dissolution rate constant at pH=9 [24]. Hence, the rate determining step of the overall dissolution reaction in the constant-rate regime at pH=12 is not the same as in the initial stage of dissolution or in the dissolution experiments at pH=9.

An explanation for this behavior can be based on the data in Figure 3, where the time-dependent evolution of dissolved silicon concentration in solution is correlated to the change in surface composition expressed in terms of the O/Si atomic ratio. Apparently, the O/Si ratio reaches a constant value of 0.2 at about the same time as the constant-rate regime for the release of silicon begins. This correlation suggests that leaching reduces the thickness of the oxidized surface layer until a minimum thickness corresponding to O/Si=0.2 (Figure 3) and pH_{iep}=8.2 (Figure 1) is reached. At this point, a dynamic equilibrium is established where dissolution (Equation 9) can only proceed if new silica-like surface oxide is formed via hydrolysis of the Si_3N_4 bulk phase (Equation 4). Hence, the experimentally observed transition at pH=12 from an initial regime with a high dissolution rate to a constant-rate regime with a very low dissolution rate is related to a change in rate-determining mechanisms. Apparently, the activation energy for hydrolysis of silicon nitride must be significantly higher than the activation energy for the dissolution of the amorphous surface oxide. This conclusion is also supported by quantum-chemical and *ab-initio* computations for hydrolysis of silicon nitride indicating that nucleophilic attack of water or hydroxyl ions leads to the formation of stable molecular complexes whose stability increases together with the degree of hydrolysis [25].

The experimental results presented here illustrate that the kinetics of silicon nitride surface oxide dissolution are controlled by two main factors, namely temperature and the activity of a nucleophilic agent (e.g. hydroxyl ions) catalyzing the hydrolysis reactions. The results further demonstrate that the extent of oxide removal through dissolution is determined by the saturation concentration of silicon species in solution and limited to the point where silicon oxide dissolution rate equals the rate of silicon nitride hydrolysis. Thus, providing sufficient silicon solubility by pH control can be used to minimize the surface oxygen concentration to a limit where O/Si=0.2 and pH_{iep}=8.2. Alternatively, soxhlet extraction treatments [13] where the powder is continuously exposed to hot, fresh solution lead to the same result as indicated by results of Biscan *et al* [26, 27].

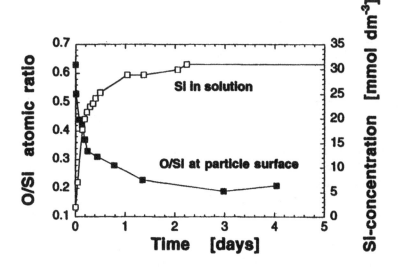

Figure 3: Si concentration in solution (□) and elemental O/Si ratio of the particle surface layer (■) obtained for a 5 vol% Si_3N_4 suspension leached at pH=12 and T=25 °C (lines are meant to guide the eye).

DEAGGLOMERATION AND DISPERSION

Fine, sub-micronsized powders are always agglomerated in the dry, as-received state. The dispersion process is strongly dependent on the strength of the interparticle bonds that keep the agglomerates together and the mechanisms that act to break the bonds. Removal or break-down of particle agglomerates is especially important considering the detrimental effect of such inhomogeneities on the mechanical properties of a sintered ceramic body. It is well known that both the size and number density of inhomogeneities play a crucial role in determining the strength and reliability of a ceramic material [3, 28].

The deagglomeration process in a moderately stirred dilute suspension of as-received silicon nitride powder (UBE SN-E10, UBE Industries, Japan) is reflected by an evolution of the particle size distribution (PSD) with time (Figure 4). When comparing the size distribution obtained with a light scattering technique to electron microscopy results (Figure 5) it becomes evident that large agglomerates of primary particles are present in the suspension; the light scattering results indicate agglomerate sizes considerably larger than the size of the primary particles or crystallites. According to the transmission electron microscopy (TEM) investigations the size of the primary particles in the silicon nitride powder fall in the range of 50 to 200 nm, whereas the agglomerates in suspension have a size of up to 200 μm. A primary particle size of 50 to 200 nm agrees well with 90 nm equivalent particle size derived from the measured specific surface area of 10.4 $m^2 g^{-1}$.

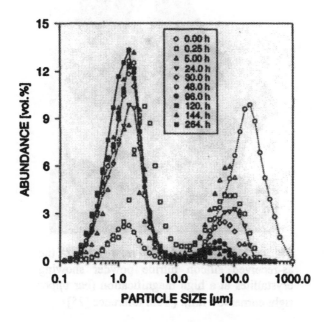

Figure 4: Evolution of particle size distribution with time in a mildly agitated 1 vol% Si_3N_4 suspension at pH≈9 and 25 °C as measured by light scattering. The initial distribution (traced out by the dotted line) is characterized by a relatively large abundance of large agglomerates that disappear at long times in favor of smaller agglomerates. (from reference [25])

Figure 4 shows that large agglomerates, with sizes between 50 and 200 μm are continuously broken down with time, and the population of smaller agglomerates (~1 μm in size) is seen to increase. The two populations consist of agglomerates that are denoted as primary and secondary agglomerates in the following discussion. The absence of peaks at intermediate sizes in the PSD during deagglomeration suggests that primary agglomerates are eroded away from the large, secondary agglomerates (i.e. primary agglomerates constitute the

Figure 5: Bright-field TEM micrograph of the as-received silicon nitride powder showing crystallites at a high magnification (bar upper right corner≈50 nm). (from reference [25])

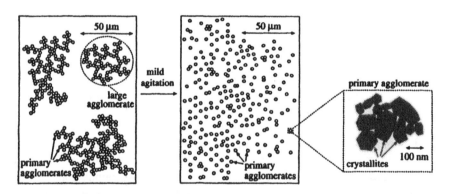

Figure 6: Schematic drawing illustrating the break-down of secondary agglomerates. (from reference [25])

building blocks of the secondary agglomerates). Hence, suspensions contain a hierarchy of agglomerates where the relative abundance of different size populations is constantly changing during the deagglomeration process. A conceptual view of this process is depicted in Figure 6.

Deagglomeration proceeds through the break-up of interparticle bonds in the aggregates when the applied force is larger than the adhesion force between aggregated particles. For a dilute suspension subjected to mild agitation, hydrodynamic drag is the dominating force on the aggregates. The magnitude of the drag force, F_d, exerted on a single particle of radius R in a flow field is approximately

$$F_d \approx 6\pi\upsilon\eta R \tag{10}$$

where υ is the fluid velocity and η the fluid viscosity. The hydrodynamic drag force required to separate two particles is proportional to the interparticle adhesion force, F_{ad},

$$F_d \approx \gamma F_{ad} \tag{11}$$

where γ is a numerical constant. As discussed previously by Desset [29], γ is close to unity when the drag force acts perpendicular to the surface and much smaller when the hydrodynamic force is parallel to the surface [29, 30]. Desset suggested $\gamma=0.01-0.1$ as a reasonable estimate for the detachment of small particles in a shear flow field [29].

For particles that are held together by attractive van der Waals forces, the adhesive interparticle force, F_{ad}, can be approximated with

$$F_{ad} \approx \frac{AR}{6D^2} \tag{12}$$

where A is the Hamaker constant and D the surface separation distance. This adhesive interparticle force is reduced by electrostatic repulsion at solution conditions where the surfaces carry a charge. The deagglomeration experiments were performed at pH$\approx$9 which is about 2.5 pH units above the isoelectric point of the silicon nitride powder (Figure 1). Therefore, it is expected that the net adhesive force between the silicon nitride particles is substantially smaller than Equation 12 predicts.

From Equations 11 and 12 one can obtain an estimate for the relative velocity between the particle and fluid phase required to separate two particles by hydrodynamic shear forces. Assuming that the maximum van der Waals attraction is reduced by one order of magnitude due to electrostatic repulsion between the charged surfaces and inserting the values $\gamma=0.01$ [29], $A=5\times10^{-20}$ J [31], and $D=0.2$ nm [3], one obtains $v\approx10^{-2}$ m s^{-1}. This value lies within the range of flow rates induced by moderate stirring, ~1-10 mm s^{-1}. Hence, it can be concluded that secondary agglomerates are held together by attractive van der Waals forces.

Apparently, the bonding forces between particles in primary agglomerates are much stronger than attractive surface forces, since primary agglomerates survive the stirring agitation (Figure 4) and even the ultrasonic treatment we used to prepare the electron microscopy samples. TEM studies revealed that agglomerates of 0.5-1 µm in size prevail in a suspension ultrasonicated at pH 10 (Figure 5).

The high strength of the primary agglomerates can be explained by the presence of rigid interparticle bridges, so-called necks. For example, measurements on boehmite [32] and silica [33] agglomerates with interparticle necks indicated agglomerate strengths of ≥20 MPa, values at least one order of magnitude higher than bond-strength estimates based on van der Waals forces only [32]. The interparticle necks in boehmite and silica were formed when the powder was dried, due to reprecipitation of dissolved species at particle contact points. Depending on the amount of dissolved material the size of the necks can be quite large (i.e. ranging from 10-100 nm).

In the case of sparingly soluble, synthetic non-oxide powders like silicon nitride, necks in the as-received powder may be formed during pyrolysis or calcination of precursor materials [3, 6] and/or through surface oxidation at particle contact points during storage. Therefore, the neck size should be on the same order as the thickness of the oxidized surface layer, i.e. approximately 1-10 nm [34-36].

A correlation between deagglomeration and dissolution is expected when interparticle necks consist of a soluble oxidic phase similar to that found on the particle surfaces. One expects that the neck size will be reduced by dissolution at a rate dictated by the dissolution kinetics. The effect of dissolution in alkaline medium on the neck size can be evaluated by using the experimentally determined dissolution rate constants. The decay of the neck radius with time, $h(t)$, is related to the dissolution rate according to

$$h(t) \approx h - \frac{k_m t}{\rho}$$

$$\text{with } k_m \left[g cm^{-2} s^{-1} \right] = 10^{-4} M_{SiN_{4/3}} k \left[mol\ m^{\pm 2} s^{-1} \right]$$

(13)

where ρ is the density of the dissolved material and $M_{SiN_{4/3}}=46.6$ g mol^{-1}.

Assuming a neck density of 2 g cm^{-3} and using the experimentally determined k-values, one finds that the neck radius should decrease by 1 nm after approximately one week at 25 °C or after only three hours at 80 °C and pH=9. At pH=12 and 25 °C the same effect is obtained after 9 hours. Thus, the decay rate is strongly temperature and pH dependent. These estimates also show that silica based necks with a radius on the order of 1–10 nm can dissolve completely over reasonably short times when pH and/or temperature are increased.

The effect of dissolution on deagglomeration was evaluated experimentally by rheological measurements using the steady-shear response of concentrated silicon nitride suspensions prepared at solution conditions corresponding to varying dissolution kinetics (Figures 7 and 8). The untreated reference suspension with a volume fraction of 35% prepared at pH 9.4 in 0.01 M NaCl-solution is shown in Figure 7. It displays a weak shear thinning behavior, with the viscosity approaching a high-shear plateau value. Subjecting the powder to a 0.01 M NaOH solution results in a significant decrease in the viscosity. The pH of this suspension decreased rapidly during sample preparation, reaching pH 9.3 after about 30 min. The viscosity at high shear rates represents the hydrodynamic resistance to flow, which is mainly controlled by the efficiency of particle packing. Hence, the reduction in viscosity suggests that the alkaline conditions result in a rapid break-up of primary agglomerates, which releases immobilized liquid and allows improved particle packing.

An attempt to obtain an even lower viscosity by subjecting the powder to extensive leaching by keeping the suspension at pH 12 for two weeks was unsuccessful. The suspension displayed a strongly shear thinning behavior (Figure 7), typical of a flocculated suspension in which large flocs are broken down by the applied shear. The flocculation was induced by the high ionic strength in this system. High counterion-concentrations, approximately 50 mM Na$^+$, were obtained because NaOH had to be continuously added to maintain the pH at 12. This effect is caused by the ionization of silicic acid at high pH. Since the silicic acid in solution is produced through powder dissolution, the amount of sodium hydroxide needed to keep a constant pH should be directly proportional to the amount of dissolved material.

In contrast, if the excess Na$^+$ and silicic acid are removed by dialysis, the degree of shear thinning and the viscosity are substantially reduced. The dialyzed suspension displays a nearly Newtonian shear behavior at 34 vol% with a high-shear viscosity comparable to a suspension viscosity at only 20 vol% measured directly after mixing. Similar results are obtained at 43 vol% solids loading (Figure 8). Hence, strong leaching leads to deagglomerated and low-viscous suspensions even at very high particle concentrations, but flocculation occurs if

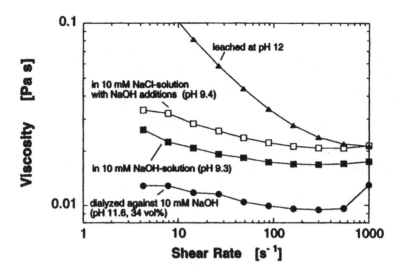

Figure 7: The steady-shear viscosity of 35 vol% suspensions at pH≈9.5 and pH≈12. (from reference [18])

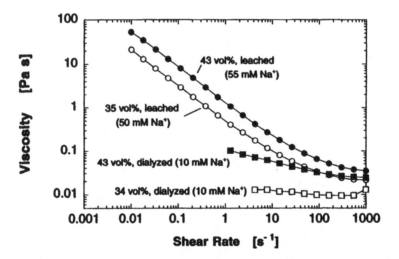

Figure 8: The steady-shear viscosity of 35 and 43 vol% suspensions at pH≈12. The approximate counter-ion concentration in solution is given in parenthesis. (from reference [18])

the corresponding increase in counterion concentration is allowed to exceed the critical flocculation concentration of the system.

INTERPARTICLE FORCES AND COLLOIDAL STABILITY

Once a deagglomerated powder is dispersed in a liquid, repulsive interparticle forces of sufficient range and magnitude are necessary to maintain colloidal stability over extended periods of time. The dominating interparticle forces in colloidal ceramic suspensions are usually van der Waals forces and electric double layer forces. The van der Waals forces are always attractive in aqueous suspensions. Double layer forces are repulsive in simple one-component systems, but may become attractive in multi-component systems. Magnitudes of both the van der Waals and the electric double layer forces are strong functions of surface separation and, for the latter, also strongly dependent on pH and ionic strength.

The force-distance relationship for particle interactions can be calculated or obtained experimentally, e.g. with atomic force microscopy. We used a colloidal-probe technique to measure forces between silicon nitride surfaces. A silicon nitride tip or a nitrided silica sphere attached to the tip of a cantilever was moved towards a flat silicon nitride wafer surface and the deflection of the cantilever due to interaction forces was recorded as a function of surface separation [37-39]. Knowing the cantilever spring constant and the diameter of the sphere, the interaction potential per unit area, $W(D)$, can be calculated by using the Derjaguin approximation

$$W(D) = \frac{F(D)}{2\pi R} \tag{14}$$

where D is the surface separation and R the probe radius. Information about the nature of the interaction forces was obtained by comparing the measurement data quantitatively with theory. Rheology combined with zeta-potential measurements provide another powerful, albeit more indirect, approach to describe the effect of dispersants on colloidal stability in ceramic suspensions. The attractive or repulsive nature of the total interaction force is reflected by suspension viscosity and shear thinning behavior. Zeta-potential measurements can help to assess double-layer force contributions to the total interaction potential. Ideally, all three techniques may be combined to link the macroscopic suspension properties to interaction forces.

van der Waals Forces

All ceramic powders experience van der Waals forces. This force is electrodynamic in origin as it arises from the interactions between oscillating or rotating dipoles within the interacting media. The van der Waals interaction free energy, $V_{vdW}(D)$, between two spheres of radius R at surface separation D, can be approximated by

$$V_{vdW}(D) = -AR/12D \qquad (15)$$

provided that $D<<R$. As seen in Equation 15, there is a direct proportionality between the magnitude of the van der Waals interaction, V, and the Hamaker constant, A. The Hamaker constant is a materials constant that depends on the dielectric properties of the two materials and the intervening media.

In the original treatment, also called the microscopic approach, the Hamaker constant was calculated from the polarizabilities and number densities of the atoms in the two interacting bodies. Lifshitz presented an alternative, more rigorous approach where each body is treated as a continuum with certain dielectric properties. This approach automatically incorporates many-body effects, which are neglected in the microscopic approach. Effective Hamaker constants for silicon nitride across various media have been calculated from optical data using Lifshitz theory. The necessary spectral parameters, the oscillator strengths and characteristic frequencies in the UV and IR range, can be determined by various means, e.g. the simple Cauchy procedure, as illustrated in Figure 9. In recent studies with a variety of different materials including silicon nitride it was demonstrated how spectroscopic ellipsometry and reflectometry could be used to measure the photon energy dependence of the refractive index, and the extinction coefficient, in the visible and near-UV range [31, 40].

The calculations by Bergström results in a Hamaker constant across air of 180 and 167 zJ for β– and amorphous silicon nitride, respectively, and a Hamaker constant across water of 55 and 48 zJ for β– and amorphous silicon nitride, respectively [31]. Like most ceramic materials, silicon nitride is thus characterized by large Hamaker constants, dominated by the dispersive interaction. The Hamaker constant estimates have been supported by direct force measurements [41]. The repulsive and attractive interactions between silicon nitride and silica across diiodomethane and 1-bromonaphthalene were measured using an atomic force microscope. We found that the sign, magnitude and separation distance scaling of the interactions in both symmetric and asymmetric systems could be well described using Hamaker constants calculated from Lifshitz theory.

Silicon-Based Structural Ceramics

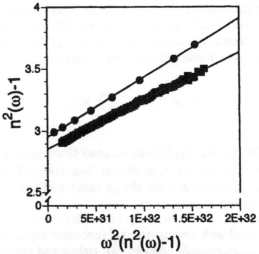

Figure 9: Cauchy plot for noncrystalline (●) and polycrystalline (■) silicon nitride.

If we want to create a colloidally stable system, some type of interparticle repulsion needs to be introduced to overcome the van der Waals attraction. Below, we will describe the two most common methods of stabilizing a colloidal suspension; either by creating an electrostatic double layer at the solid-liquid interface or by adsorbing polymers or surfactants on the particle surfaces.

Electrostatic Double-Layer Forces and Charge Regulation

The interaction between two charged particles in a polar medium is related to the osmotic pressure created by the increase in ion concentration between the particles where the electrical double layers overlap. The repulsion can be calculated by solving the Poisson-Boltzmann (PB) equation, which describes the potential and ion concentration in two overlapping double layers,

$$\frac{d^2\Psi}{dx^2} = -\frac{e}{\varepsilon\varepsilon_0}\sum_i \rho_{0,i}z_i \exp\left(\frac{-z_i e\Psi_x}{kT}\right) \qquad (16)$$

where Ψ_x is the potential at distance x and ρ_0 is the bulk concentration of ions of valency z.

Considering the chemical equilibria that are responsible for the development of charges at surface sites, it is seen that the charge density, which determines the potential distribution in the diffuse layer itself, is a function of the surface potential. The thermodynamic background of charge regulation for silicon nitride has been described by Zhmud and Bergström [42]. The dissociation reactions for the potential determining reactions are described in Equation 1 with the potential dependent effective equilibrium constants given by

$$K_i = K_i^0 \exp(-u_s \sum_j \upsilon_{ij} z_j) \tag{17}$$

where K_i^0 is the intrinsic equilibrium constant characterizing the ith equilibrium at zero potential, u_s is the dimensionless surface potential, and υ_{ij} and z_j are the stoichiometric coefficient and the charge number of the jth component of the ith reaction, respectively.

The solution to the full sets of equations (described in detail in reference [38]) has been evaluated with a computational procedure based on two nested iteration cycles. Force measurements between two etched and two plasma treated silicon nitride surfaces consisting of a tip and a flat substrate of the same chemical composition were conducted in an aqueous electrolyte solution in the pH range 3-11 [38]. The silica system consisted of a glass sphere attached to a tip–less cantilever and an oxidized silica plate. From the experimental force curves, a number of measurement points are chosen and organized in triplets as force-separation-pH. These points serve as the basis for obtaining surface ionization parameters, which are evaluated by minimization of the deviation between the experimental and calculated force curves. Fixed values were assigned to the association constant and the Hamaker constant. Site densities and protonation constants of the Si_2NH and $SiOH$ groups, and the tip radius were used as fitting parameters and assigned relevant starting values.

Experimental force curves are shown in Figure 10 for plasma cleaned silicon nitride surfaces. The force is purely attractive at pH 3.5 but an increasing repulsion is seen at successively higher pH. The van der Waals force is completely overcome by electrostatic repulsion above pH 9. The low isoelectric point indicates that the surfaces are heavily oxidised. Force curves between etched silicon nitride surfaces (Figure 11) show that the isoelectric point has been shifted to a much higher pH due to a decrease in silanol group surface density.

Fitting experimental force curves to DLVO-theory using the boundary conditions of constant charge or constant potential yields values of surface charge and surface potential that are independent of separation distance. The numerical procedure employed here computes the distance dependence of both surface charge and surface potential. Figure 12 shows potential and charge density for both silicon nitride systems and the silica system at intersurface distances of 1 nm.

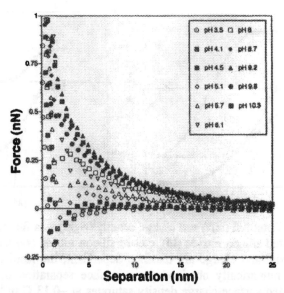

Figure 10: Force measurement as a function of separation and pH in 1 mM KCl between two identical, amorphous, plasma cleaned silicon nitride surfaces. (from reference [38])

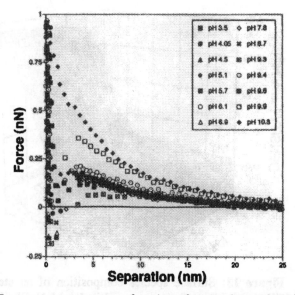

Figure 11: Force measurement as a function of separation and pH in 1 mM KCl between two identical, amorphous, etched silicon nitride surfaces. (Ref. [38])

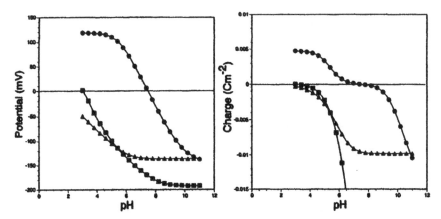

Figure 12: Potential (left) and charge density (right) as a function of pH for plasma treated silicon nitride (■), etched silicon nitride (●) and silica (▲). Both potential and charge are distance dependent; the results shown here have been numerically obtained for a surface separation of 1 nm. The plasma treated surface charge density saturates at -0.13 C m^{-2} (one charge per 1.2 nm^2) at high pH.

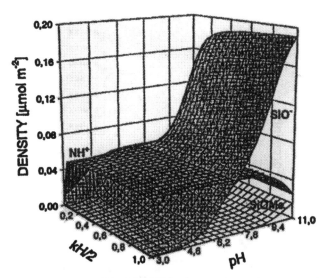

Figure 13: Surface species composition of an etched silicon nitride surface, measured in 1 mM NaCl. (from reference [38])

Note how the isoelectric point for plasma treated silicon nitride surfaces is closer to silica than to etched silicon nitride, with similar charge densities at low pH.

Densities of the surface species for the etched silicon nitride system are shown in Figure 13. Here the isoelectric point (or the point of zero charge) can be seen at the intersection of $Si_2NH_2^+$ and SiO^- density plots. Another observation at extreme pH and small separations, is the formation of ion pairs between surface groups and electrolyte ions in solution.

Electrosteric Stabilization

Water-soluble anionic and cationic polymers (i.e. polyelectrolytes) are commonly used as dispersants for ceramic powders in aqueous media [1]. The polyelectrolyte adsorbs at the solid-liquid interface and infers a repulsive force between particles, which keeps the particles well dispersed. Polyelectrolyte adsorption is highly dependent on the electrostatic interactions between the polyelectrolyte and the surface; hence, the surface chemistry of the solid phase and the solution properties of the polyelectrolyte are important parameters, regulated by the pH and the ionic strength [43]. The pH controls the sign and density of charges on the surface, and the degree of dissociation and the conformation in solution of a weak polyelectrolyte. The salt concentration also affects the conformation of the polyelectrolyte in solution and the screening of the electrostatic interaction between charged polymer segments and the surface charges.

The term electro-steric stabilization is often used to describe how polyelectrolytes act as dispersants. Electrosteric stabilization is a combination of an electrostatic double-layer repulsion and a steric repulsion, where the relative importance of the respective contributions is closely related to the polymer segment density profile at the interface. If the polyelectrolyte adsorbs in a flat conformation, the steric repulsion is short-range, and the stabilization mechanism is mainly electrostatic. This is usually the case when the polyelectrolyte is highly charged, having an extended conformation, and the particle surface is oppositely charged. With thicker adsorbed layers, having chains protruding into the solution, the steric contribution will become more important. However, there is always an electrostatic contribution since the adsorption of a highly charged polyelectrolyte on a weakly charged, amphoteric oxide surface usually results in an increase of the net surface charge density.

We have investigated the surface forces induced by adsorption of polyacrylic acid (PAA) onto silicon nitride with a $pH_{iep} \approx 4.7$. AFM force-distance curves of systems containing background electrolyte and polyacrylic acid are shown in Figure 14 where force data is normalized with the probe radius, R. The surfaces experience a long-range repulsion followed by attraction at close separations. At the point where the force gradient of the attractive van der Waals force exceeds the cantilever spring constant (i.e. at a surface separation of ~4 nm), the probe jumps into contact. The applied force has to be increased further to achieve hard wall contact, suggesting a thin (<2 nm) adsorbed layer of PAA. The inset in Figure 14 includes a fit with DLVO-theory illustrating good agreement between the calculated total ionic strength, I, and the measured Debye length.

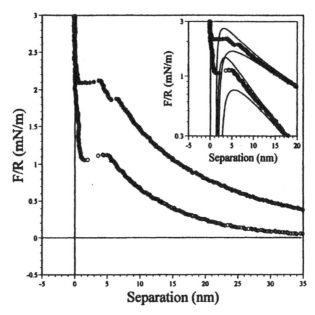

Figure 14: Normalized forces measured between a spherical Si_3N_4 probe and a Si_3N_4 flat surface after adsorption from a 500 ppm PAA-50000 solution (I=1.1×10^{-3}M) at pH 5.5 (O), and after subsequent flushing with 0.1×10^{-3}M NaCl at pH 8.5 (●). Lines in the inset are fits with DLVO-theory using constant charge and constant potential boundary conditions. The fitted potentials are $|\psi_0|$=50 mV and $|\psi_0|$=95 mV, respectively. (from reference [39])

The extent of PAA adsorption was estimated from the surface potentials extracted by fitting the force data with DLVO-theory. A fitted potential of $|\Psi_0|$=50 mV after PAA adsorption was compared to $|\Psi_0|$=30 mV of silicon nitride in pure electrolyte at equal pH=5.5. The relative small increase in surface potential suggests that only low amounts of PAA were adsorbed. This is not unexpected since adsorption isotherms for PAA adsorption onto silicon nitride powders also showed low adsorbed amounts (i.e. on the order of 0.05 mg m^{-2} at pH>pH$_{iep}$) [44]. The low affinity isotherms can be attributed to the electrostatic repulsion between the negatively charged surface and polyelectrolyte.

The adsorbed PAA layer attains a comparably flat conformation at the solid/liquid interface. Comparing the layer thickness of <2 nm with the root-mean-square end-to-end distance ($\sqrt{R_0^2} \approx$ 5-10 nm for M$_w$=10000) [45] of the polymer in solution suggests that a substantial reconformation of the polymer chains at the interface occurs upon adsorption. It appears that the electrostatic repulsion between surface and polyelectrolyte causes a very slow adsorption kinetics as pointed out previously by Cohen Stuart *et al* [46]. This gives the adsorbing macromolecules time to relax into a flat interfacial conformation.

Force-distance curves were also measured at a higher total ionic strength (Figure 15) to investigate the importance of electrostatic interactions on PAA adsorption. The adsorption from a 2000 ppm PAA-10000 solution at pH=5.5 (I=19 mM) leads to force curves exhibiting a weak electrostatic interaction with an exponential decay at 10-15 nm separation, followed by a van der Waals attraction. The inset in Figure 15 shows that the Debye length corresponding to the calculated ionic strength agrees well with the long-range tail of the force curve. A polymeric (steric) force is detected at surface separations smaller than 7 nm. Compressing the adsorbed PAA-10000 layer reveals a layer thickness of approximately 3-4 nm, which is approximately twice as thick as the PAA-50000 layer adsorbed at low ionic strength. This supports the notion that screening the electrostatic repulsion by increasing the ionic strength results in a higher adsorbed amount of PAA. However, the range of the steric force is still quite small, which implies a relatively flat PAA layer.

Figure 15 also displays the force-distance relationship for a system where PAA was adsorbed from a solution containing 0.1 mM Y^{3+} and 20.5 mM NaCl ($I\approx$22 mM). Similar to the measurement with a high ionic strength solution that did not contain yttrium, the interaction curve exhibits a monotonic repulsive force at all distances upon compression. The total ionic strength, I, is comparable in both measurements with and without the addition of Y^{3+}, but the decay length at separation distances >10 nm is much larger in the latter case and clearly not attributive to an electrostatic force (see inset in Figure 15). However, after flushing with pure NaCl electrolyte solution (I=0.1 mM) at pH 5.5 the repulsive force can be fitted to DLVO-theory and a jump into contact occurs at ~4 nm separation distance (Figure 15). Hence, it appears that the long-range, non-electrostatic repulsive force seen for the sample with Y^{3+} content is caused by weakly bound polymer aggregates that can be removed by flushing with new solution. The sample with yttrium and a high total ionic strength of I=22 mM further exhibited a very long-range adhesion force. This example emphasizes the detrimental effect of multivalent cations on colloidal stability of ceramic suspensions with anionic dispersants.

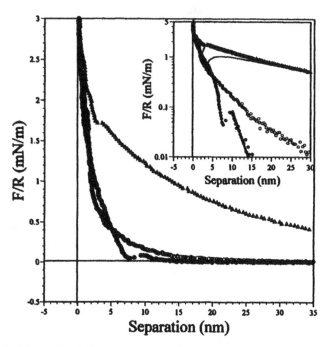

Figure 15: Normalized forces measured between a Si_3N_4 probe and a Si_3N_4 flat surface after adsorption at pH 5.5: 2000 ppm PAA-10000 solution ($I = 1.9 \times 10^{-2}$ M) (●), 500 ppm PAA-50000 solution ($I = 2.2 \times 10^{-2}$ M) containing 0.1×10^{-3} M YCl_3 (○) and after subsequent flushing with 0.1×10^{-3} M NaCl at pH 5.5 (▲). Lines in the inset are fits with DLVO-theory using constant charge and constant potential boundary conditions (▲). The fitted potential is $|\psi_0| = 90$ mV. The thick solid line represents the decay length ($\kappa^{-1} = 2.2$ nm) of the electrostatic potential for a 1:1 electrolyte at 1.9×10^{-2} M concentration. (from reference [39])

At first sight, it seems that the qualities of PAA as a dispersant may be rather limited with respect to colloidal processing of silicon nitride, which is usually done at alkaline conditions. Only a moderate increase in surface potential is observed upon adsorption and adsorbed layer thicknesses are too thin to provide an effective steric repulsion. Rheological measurements also indicate that PAA additions can afford silicon nitride stabilization at moderate solids loading only in the range $pH < pH_{iep} + 1.5$ [47]. At higher pH, PAA adsorption becomes negligible due to electrostatic repulsion between polymer and Si_3N_4 surface. As discussed before, concentrated suspensions with a low viscosity can be obtained without any dispersant at all when proper deagglomeration and pH-control are provided. However, the main benefit of PAA may be its role as a dispersant for many sintering additives such as MgO and Al_2O_3, i.e. PAA additions are necessary to effectively minimize heterocoagulation of different slurry components at alkaline pH.

It has been reported that PAA stabilizes some ceramic materials even at a pH above pH_{iep}. As discussed previously in more detail [39], material-dependent non-electrostatic interactions between polyelectrolyte segments and adsorption sites at the particle surface might be the reason for the observed fundamental differences in PAA adsorption onto Si_3N_4 and, for example, Al_2O_3 or ZrO_2. For the latter case it is possible that the formation of irreversible surface-segment bonds at the solid-liquid interface promote an extended polyelectrolyte conformation and, thus, a pronounced (electro-)steric interparticle repulsion. Apparently, no such irreversible surface-segment bonds are formed in the process of PAA adsorption onto Si_3N_4, which allows PAA macromolecules to adsorb in a flat conformation.

SUMMARY

Oxidation of the thermodynamically unstable silicon nitride in contact with air or water leads to formation of an amorphous silica-like oxide layer on the powder surface. The surface chemistry of the powder depends on the degree of oxidation, because the relative amount of potential determining secondary amine and silanol surface groups is altered by oxidation. Dissolution of the oxide layer in an aqueous medium leads to the release of silicon and, therefore, also affects the oxide layer composition and the powder surface chemistry. The dissolution behavior of the surface oxide at alkaline pH resembles the dissolution of amorphous silica, both in terms of activation energy of the overall dissolution reaction and the rate dependence on pH and temperature. In particular, dissolution rates are increased by up to three orders of magnitude, if the pH and/or temperature are increased.

Extensive leaching at high pH and/or elevated temperature causes a nearly complete removal of the oxide layer, a situation where dissolution does not continue unless the silicon nitride surface is reoxidized via hydrolysis. This dynamic equilibrium of surface reactions is characterized by a surface composition with a minimum atomic O/Si ratio of 0.2 and an isoelectric point of $pH_{iep} = 8.2$.

Deagglomeration of a fine silicon nitride powder under mild agitation proceeds in two stages. The relatively rapid decay of the large, secondary agglomerates is dominated by the break-up of adhesive particle-particle bonds originating from attractive surface forces. The slower break-up of hard, relatively small primary agglomerates (~1 μm) could be related to the continuous weakening of interparticle necks due to oxide dissolution. It was found that the magnitude of the imposed hydrodynamic drag forces determines the rate of deagglomeration during the early stage, whereas dissolution processes take over as the rate determining mechanism during the later stage.

The high strength of interparticle necks in primary agglomerates implies that necks either have to be dissolved or broken by excessive mechanical forces to obtain a deagglomerated suspension. In a concentrated suspension subjected to high energy milling inertial forces transmitted through collisions with other clusters or the milling media become important and make the deagglomeration process a combination of cluster erosion and attrition. Although intense milling often is the solution, optimization of the dissolution rate may solve a deagglomeration problem in a shorter time with much smaller energy consumtion. For the oxidized silicon nitride system, rapid dissolution is obtained at high pH at elevated temperatures. It is also necessary, however, to optimize the peptization process with regard to the amount of added base. Attempting to achieve deagglomeration at room temperature solely by neck dissolution requires larger amounts of base and the associated increase in ionic strength can lead to flocculation.

The surface forces in silicon nitride systems have been described. The Hamaker constants in various media have been estimated and compared to direct measurements. A novel procedure adapted to obtain surface composition, potential and charge information from AFM force measurements – evaluated with a numerical computation using charge regulation – has been described. This method has some interesting applications; it can be used as a complementary technique to pH-potentiometric titration for surface charge determination where the results reflect the actual surfaces and can thus be used for simulation of force profiles and colloidal stability.

By using a novel probe for direct measurement of forces between silicon nitride surfaces, it could be shown that polyacrylic acid, a commonly used anionic dispersant, adsorbes in a flat conformation onto silicon nitride at pH>pH$_{iep}$. The adsorbed amount can be increased by increasing the ionic strength, but adsorbed layers are still rather flat. Hence, the contribution of adsorbed PAA to the stabilization mechanism of silicon nitride is purely electrostatic.

ACKNOWLEDGEMENTS

This work was performed within "The Brinell Centre - Inorganic Interfacial Engineering", supported by the Swedish National Board for Industrial and Technical Development (NUTEK) and the following industrial partners: Atlas

Copco Secoroc AB, Erasteel Kloster AB, Höganäs AB, Sandvik AB and Seco Tools AB.
The authors would like to emphasize the importance of collaborations with many other scientists in the course of this work. We are especially grateful to Boris Zhmud, Joseph Yanez, Sabine Desset and Marie Ernstsson for their valuable contributions.

REFERENCES

[1] R.J. Pugh and L. Bergström (eds), *Surface and colloidal chemistry in advanced ceramics processing*, Marcel Dekker, New York, 1994.

[2] W.B. Russel, D.A. Saville and W.R. Schowalter, *Colloidal Dispersions*, University Press, Cambridge 1989.

[3] F.F.Lange, "Powder processing science and technology for increased reliability", J. Am. Ceram. Soc., 72 3-15 (1989).

[4] I.A. Aksay, "Microstructure control through colloidal consolidation"; pp. 94-104 in *Advances in Ceramics*, Vol. 9, Edited by J.A. Mangels and G.L. Messing. American Ceram. Soc., Columbus, 1984.

[5] L. Bergström and E. Bostedt, "Surface Chemistry of Silicon Nitride Powders: Electrokinetic Behaviour and ESCA Studies", *Colloid Surf.*, 49 183-197 (1990).

[6] T. Yamada, "Preparation and Evaluation of Sinterable Silicon Nitride Powder by Imide Decomposition Method", *Am. Ceram. Soc. Bull.*, 72 99-106 (1993).

[7] M. Peuckert and P. Greil, "Oxygen distribution in silicon nitride powders"; *J. Mater. Sci.*, 22 3717-3720 (1987).

[8] H. Stadelmann, G. Petzow, and P. Greil, "Effects of Surface Purification on the Properties of aqueous Silicon Nitride Suspensions"; *J. Eur. Ceram. Soc.*, 5 155-163 (1989).

[9] L. Bergström, M. Ernstsson, B. Gruvin, R. Brage, B. Nyberg, and E. Carlström, "The Effect of Wet and Dry Milling on the Surface Properties of Silicon Nitride Powders"; pp. 1005-1014 in *Ceramics Today - Tomorrow's Ceramics*, Edited by P. Vincenzini. Elsevier Science Publishers B.V., Amsterdam, 1991.

[10] P. Greil, R. Nitzsche, H. Friedrich, and W. Hermel, "Evaluation of Oxygen Content on Silicon Nitride Powder Surface from the Measurement of the Isoelectric Point"; *J. Eur. Ceram. Soc.*, 7 353-359 (1991).

[11] H. Lange, G. Wötting, and G. Winter, "Silicon Nitride - From Powder Synthesis to Ceramic Materials"; *Angew. Chem. Int. Edition*, 30 1579-1597 (1991).

[12] S. Natansohn, A.E. Pasto, and W.J. Rourke, "Effect of Powder Surface Modifications on the Properties of Silicon Nitride Ceramics"; *J. Am. Ceram. Soc.*,

76 2273-2284 (1993).

[13] V.A. Hackley, P.S. Wang, and S.G. Malghan, "Effects of soxhlet extraction on the surface oxide layer of silicon nitride powders"; *Mater. Chem. Phys.*, **36** 112-118 (1993).

[14] W. Dressler and R. Riedel, "Progress in Silicon-Based Non-Oxide Structural Ceramics"; *Int. J. Refract. Hard Mater.*, **15** 13-47 (1997).

[15] E. Laarz, G. Lenninger, and L. Bergström, "Aqueous Silicon Nitride Suspensions: Effect of Surface Treatment on the Rheological and Electrokinetic Properties"; *Key Eng. Mater.*, **132-136** 285-288 (1997).

[16] S.I. Raider, R. Flitsch, J.A. Aboof, and W.A. Plisken, "Surface Oxidation of Silicon Nitride Films"; *J. Electrochem. Soc.*, **123** 560-565 (1976).

[17] B.V. Zhmud, and L. Bergström, "Dissolution Kinetics of Silicon Nitride in Aqueous Suspension"; *J. Colloid Interface Sci.*, **218** 582-584 (1999).

[18] E. Laarz and L. Bergström, "Deagglomeration, dissolution and stabilization of silicon nitride in aqueous media"; pp. 159-166 in *Ceramic Processing Science VI*, Edited by S. Hirano, G.L. Messing, and N. Claussen. The American Ceramic Society, Westerville, 2001.

[19] R.K. Iler, *The Surface Chemistry of Silica*, John Wiley & Sons, New York, 1979.

[20] J.D. Kubicki, Y. Xiao, and A.C. Lasaga, "Theoretical reaction pathways for the formation of $[Si(OH)^5]^{-1}$ and the deprotonation of orthosilicic acid in basic solution"; *Geochim. Cosmochim. Acta*, **57** 3847-3853 (1993).

[21] R.L. Segall, R.S.C. Smart, and P.S. Turner, "Oxide Surfaces in Solution"; pp 527-576 in *Surface and Near-Surface Chemistry of Oxide Materials*, Edited by J. Nowotny and L.C. Dufour, Elsevier Science Publishers B. V., Amsterdam, 1988.

[22] Y. Xiao and A.C. Lasaga, "Ab initio quantum mechanical studies of the kinetics and mechanisms of quartz dissolution: OH-catalysis"; *Geochim. Cosmochim. Acta*, **60** 2283-2295 (1996).

[23] Y. Xiao and A.C. Lasaga, "Ab initio quantum mechanical studies of the kinetics and mechanisms of silicate dissolution: $H^+(H_3O^+)$ catalysis"; *Geochim. Cosmochim. Acta*, **58** 5379-5400 (1994).

[24] E. Laarz, "Colloidal Processing of Non-Oxide Ceramic Powders in Aqueous Medium", PhD thesis, Royal Institute of Technology (KTH), Department of Materials Science and Technology, Stockholm, 2000.

[25] E. Laarz, B.V. Zhmud and L. Bergström, "Dissolution and Deagglomeration of Silicon Nitride in Aqueous Medium", *J. Am. Ceram. Soc.*, **83** 2394-2400 (2000).

[26] N. Kallay, J. Biscan, T. Smolic, S. Zalac, and M. Krajnovic, "Potentiometric Determination of Reaction Enthalpies at Silicon Nitride/Water Interface"; *Polish J. Chem.*, **71** 594-602 (1997).

[27] J. Biscan, N. Kallay, and T. Smolic, "Determination of iso-electric point

of silicon nitride by adhesion method"; *Colloid Surf. A*, **165** 115-123 (2000).

[28] E. Carlström, "Surface and Colloid Chemistry in Ceramics: An overview"; pp 1-28 in *Surface and Colloid Chemistry in Advanced Ceramics Processing*, surfactant science series Vol. 51, Edited by R.J. Pugh and L. Bergström, Marcel Dekker: New York, 1994.

[29] S. Desset, "Redispersion of Alumina Particles in Water: Influence of the Surface State", *PhD thesis*, L'Universite Paris 6, Paris, 1999.

[30] M.M. Sharma, "Factors controlling the hydrodynamic detachment of particles from surfaces"; *J. Colloid Interface Sci.*, **149** 121-134 (1992).

[31] L. Bergström, "Hamaker constants of inorganic materials"; *Advances in Colloid and Interface Science*, **70** 125-169 (1997).

[32] S. Kwon and G.L. Messing, "The Effect of Particle Solubility on the Strength of Nanocrystalline Agglomerates: Boehmite"; *NanoStruct. Mater.*, **8** 399-418 (1997).

[33] A. Maskara and D.M. Smith, "Agglomeration during the Drying of Fine Silica Powders, Part II: The Role of Particle Solubility"; *J. Am. Ceram. Soc.*, **80** 1715-1722 (1997).

[34] S.M. Chu, "The dispersion of silicon nitride in aqueous media"; *M.Sc. thesis*, MIT, 1990.

[35] M.N. Rahaman, Y. Boiteaux, and L.C. de Jonghe, "Surface Characterization of Silicon Nitride and Silicon Carbide Powders"; *Am. Ceram. Soc. Bull.*, **65**, 1171-1176 (1986).

[36] P.S. Wang, S.M. Hsu, S.E. Malghan, and T.N. Wittberg, "Surface Oxidation-Kinetics of Si_3N_4-4-Percent-Y_2O_3 Powders Studied by Bremsstrahlung-Excited Auger-Spectroscopy", *J. Mater. Sci.*, **26** 3249-3952 (1991).

[37] A. Meurk, "Force measurements using scanning probe microscopy", PhD thesis, Royal Institute of Technology (KTH), Department of Materials Science and Technology, Stockholm, 2000.

[38] B.V. Zhmud, A. Meurk and L. Bergström, "Evaluation of surface ionization parameters from AFM data," *J. Colloid Interface Sci.*, **207** 332-343 (1998).

[39] E. Laarz, A. Meurk, Y.A. Yanez, and L. Bergström, "Silicon nitride colloidal probe measurements: Interparticle forces and the role of surface-segment interactions in poly(acrylic acid) adsorption from aqueous solution", *J. Am. Ceram. Soc.*, **84** 1675-1682 (2001).

[40] R.H. French, R.M. Cannon, L.K. DeNoyer, and Y.M. Chiang, "Full spectral calculation of non-retarded Hamaker constants for ceramic systems from interband transition strength", *Solid State Ionics*, **75** 13-33 (1995).

[41] A. Meurk, P.F. Luckham and L. Bergström, "Direct measurement of repulsive and attractive van der Waals forces between inorganic materials", *Langmuir*, 13 3896-3899 (1997).

[42] B.V. Zhmud and L. Bergström, "Charge regulation at the surface of a porous solid", pp. 567-592 in *Surfaces of Nanoparticles and Porous Materials*, Edited by J.A. Schwarz and C. Contescu, Marcel Dekker, Inc, New York, 1999.

[43] G.J. Fleer, M.A. Cohen Stuart, J.M.H.M. Scheutens, T. Cosgrove, and B. Vincent, *Polymers at Interfaces*, Chapman & Hall, London, 1993.

[44] V.A. Hackley, "Colloidal Processing of Silicon Nitride with Poly(acrylic acid): I, Adsorption and Electrostatic Interactions"; *J. Am. Ceram. Soc.*, 80 2315-2325 (1997).

[45] W.J. Walczak, D.A. Hoagland, and S.L. Hsu, "Spectroscopic Evaluation of Models for Polyelectrolyte Chain Conformation in Dilute Solution"; *Macromolecules*, 29 7514-7520 (1996).

[46] M.A. Cohen Stuart, C.W. Hoogendam, and A. de Keizer, "Kinetics of polyelectrolyte adsorption"; *J. Phys.: Condens. Matter*, 9 7767-7783 (1997).

[47] E. Laarz and L. Bergström, "The effect of anionic polyelectrolytes on the properties of aqueous silicon nitride suspensions", *J. Eur. Ceram. Soc.*, 20 431-440 (2000).

VISCOELASTIC PROPERTIES OF CONCENTRATED SILICON NITRIDE SLURRIES

Xue-Jian Liu, Mei-Fang Luo*, Xi-Peng Pu, Li-Ping Huang, Hong-Chen Gu*
Shanghai Institute of Ceramics, Chinese Academy of Sciences, Shanghai 200050, China
*East China University of Science & Technology, Shanghai 200237, China

ABSTRACT

The colloidal behavior, rheological behavior, and viscoelastic behavior of concentrated Si_3N_4 slurries are investigated in detail by electrophoresis and rheological measurements in order to prepare a suspension with low viscosity and high solid content. The X-ray photoelectron spectroscopy reveals a decrease of the surface density of acidic silanol group after that Si_3N_4 powder is surface modified. The isoelectric point of the modified Si_3N_4 particle shifts to basic region gently and the zeta potential in alkaline region evidently increases in magnitude. Attempts have been made to apply rheological models to the suspensions with various solid volume fractions (ϕ). For the as-received suspensions, the Sisco model provides the best fit in the range of ϕ 0.30 while the Casson model in 0.35 ϕ 0.45. The shear behavior of modified suspensions fits to Sisco model in the range of ϕ 0.40 and Casson model in 0.45 ϕ 0.54. The rheological behavior of modified suspensions is improved efficiently. The critical strain decreases and the linear viscoelastic regime narrows continuously with increasing solid concentration. With increasing solid concentration, the magnitude of the viscoelastic response increases, the suspension transforms from more viscous to more elastic progressively and the characteristic frequency shifts toward lower frequencies. The viscoelastic behaviors can be understood in terms of the structural relaxation time of the suspension.

1. INTRODUCTION

Colloidal processing technique such as slip casting or pressure slip casting has commonly been accepted as one of the most attractive routes in the fabrication of ceramic components with complex shapes. This is because that these techniques offer improved homogeneity and reliability, which can hardly be achieved through the conventional alternatives, e.g. dry pressing [1-2]. In colloidal processing, water has always been used as a dispersing media because of low-cost and environmental protection. For this purpose, ceramic powder should be dispersed effectively in water medium to form a desirable suspension with the low viscosity and high solid content which will produce a better green body as well as fine microstructure [1-2]. This can usually be achieved with the help of appropriate dispersants that stabilize ceramic particles via an electrostatic repulsion or steric hindrance, or a combination of both.

The dispersion of colloidal particles in an aqueous media by electrostatic and/or steric effect has been investigated extensively [3]. Both attractive and repulsive forces between particles depending on surface charges manage the former. The net effect of these forces determines the state of dispersions. Ceramic particles are naturally positively or negatively charged depending on the pH conditions. The amount of charge can be easily measured via zeta potential. Zeta potential is the potential at the shear plane and is generally accepted as a reasonable measure of the amount of surface charge. The point at which the zeta potential is zero is termed the isoelectric point (IEP). At this pH value the electrostatic repulsive force is zero, so the particles aggregate together in a purely electrostatically stabilized system [4-5]. However, for large zeta potential in magnitude electrostatic repulsive force is larger than van der Waals attractive force, which prevents particles from flocculating. Hence, during preparing a stable ceramic suspension it is necessary to have the knowledge of the IEP, so that a system has a pH far from that of IEP. In general, an electrostatic stabilization of suspension can be achieved by manipulating electrostatic charges on the particle surface by controlling pH and/or adding a dispersant into the suspension, which is adsorbed on the surface of particles thus increasing the repulsive force [3, 6].

In all types of suspension shape-forming techniques ranging from slip casting to injection molding, the rheological properties of concentrated suspension play a key role in controlling the shape-forming behavior and optimizing the properties of the green body. For example, during slip casting of a suspension, the optimized rheological properties will minimize the density gradients, mass segregation of different components, and avoid strain recovery of the cast body after consolidation [7-10]. Fundamentally, the rheological properties of concentrated colloidal suspensions are determined by the interplay of thermodynamic and fluid mechanical interactions. With particles in colloidal size range (at least one dimension < $1 \mu m$), the range and magnitude of the interparticle forces will have a profound influence on the suspension structure and, hence, the rheological behavior [10-11].

The rheological behavior of concentrated suspension is affected by a number of factors including particle size distribution, particle shape and volume fraction of solid. The flow curve of suspension can provide information that relates to the interactions between the particles and the media. In particular, if an appropriate model can represent the data, the evaluation may become more convenient and effective [5]. Several models have been developed for non-Newtonian systems, including Bingham plastic model, Casson model, and Herschel-Buckley model [12]. These models have been widely and successfully used to explain, characterize, and predict the flow behavior for various systems. Nevertheless, not much has been reported in regard to the study of ceramic suspension systems by means of these models. In addition, colloidal suspensions usually exhibit viscoelastic properties that become more and more significant with increasing particle concentration. Thus, in order to achieve a through characterization of the rheological behavior of concentrated ceramic suspensions, the viscoelastic response should be taken into consideration, also because it can provide more detailed information about the structural conditions of the disperse phase in the equilibrium state.

The present work is intended to discuss how rheological measurements can elucidate important information concerning the properties of ceramic suspensions. We focused our attention mainly on the rheological characterization of aqueous Si_3N_4 suspensions. The influence of the surface modification of Si_3N_4 powders on the colloidal behavior, rheological behavior, and viscoelastic behavior of suspension is systematically investigated. Additionally, attempts have been made to apply rheological models to the suspensions.

2. EXPERIMENTAL PROCEDURE
2.1 Materials and Surface Modification

Commercially available Si_3N_4 powder (Starck LC-12, Germany) with an average particle size of $0.60\mu m$ and a BET specific surface area of $21.0\ m^2 \cdot g^{-1}$ was used for the present study. In order to adjust the surface properties, the powders was leached with acidic aqueous solution (pH=2) at first and then repeatedly rinsed with deionized water until that no change in conductivity was detected [13]. The dispersant used was an inorganic compound which was the best dispersing agent selected from a number of commercial dispersants through a series of sedimentation tests [14].

2.2 Preparation and characterization of suspensions

Suspensions of different solid volume fraction (ϕ) for rheological measure were prepared by the successive addition of various amounts of Si_3N_4 powders into deionized water containing 1.0wt% dispersant, as calculated on the basis of dry Si_3N_4 powder. The solution pH was preadjusted to yield a value close to target pH (pH=11), whereas subsequent adjustment of suspension pH was made using tetramethylammonium hydroxide (TMAH) or HCl. Wetting and dispersion were improved by both intermittent ultrasonic treatment and continuous stirring. This procedure was followed by ball milling in a ceramic jar with Si_3N_4 balls.

Zeta potential of Si_3N_4 particles was measured with dilute aqueous suspension in 10^{-3}M NaCl electrolyte on Zetasizer 4 (Malvern, UK). To determine the zeta potential as a function of pH, 10^{-2}N HCl and 10^{-2}N NaOH solutions were used to adjust pH to the desired values.

Rheological characterization was performed on Rheometeric Fluid Spectrometer II at 25 °C, with a couette (cup radius: 34.0mm, bob radius: 32.0mm, bob length: 33.3mm) or a cone and plate measurement geometry (radius: 12.5mm, angle: 0.1rad). A small amount of silicone oil was floated on the sample to prevent evaporation. Steady shear measurements of the examined systems were performed by incrementing shear rate and measuring apparent viscosity. The viscoelastic properties of the suspensions were examined under oscillatory shear conditions. In order to determine the limits of linear viscoelastic regime, strain sweep experiments from 0.01 to 50 percent of strain were carried out at constant frequency. The mechanical spectra of the examined suspensions were evaluated by applying frequency sweep experiments in the linear viscoelastic region.

3. RESULTS AND DISCUSSION
3.1 Colloidal Behavior of Si_3N_4 Suspension

The effect of surface modification on zeta potential of Si_3N_4 particles is shown in Fig. 1. As can be seen, the IEP of modified Si_3N_4 particles shifts to alkaline region gently and the zeta potential of modified Si_3N_4 particle in alkaline region evidently increases in magnitude. According to Bergstrom [15], the shift of IEP is resulted from the decrease of the relative site density of acidic silanol group (-SiOH). The X-ray photoelectron spectroscopy (XPS) reveals a decreased O/Si ratio and an increased N/Si ratio, i.e., a decrease of the surface density of acidic silanol group after that the powder is leached [16]. This may be resulted from the removal of an amorphous oxide phase by leaching solution, thus leaving a surface with relatively more basic groups. Denny et al. showed that metal cation impurities such as Al^{3+} and Fe^{3+} in minute amounts would significantly depress the dissolution of silica probably because of the formation of a stable complex between silanol group and polyvalent metal cation [17]. This process can be described as following reaction:

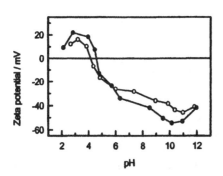

Fig.1 Effect of surface modification on zeta potential of Si_3N_4 powder (●)surface modified, (○) as received.

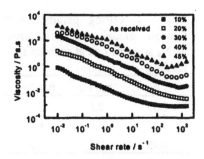

Fig. 2 Viscosity curves of as-received
Si₃N₄ suspensions at various solid
loading.

$$(-SiOH)_m + Fe^{3+} = (-SiOH)_{m-n}(-SiO)_n Fe^{3-n} + n\,H^+ \qquad (m \geq n, 3 \geq n \geq 1)$$

The results of inductively coupled plasma (ICP) analysis show that the amount of
ferric oxide (Fe_2O_3) in modified powder decreased by 47% over the as-received
powder. This indicates that acid leaching clears away metal impurities and thereby
accelerates the dissolution of silica at the surface of Si_3N_4 powder.

3.2 Rheological Behaviors of Si₃N₄ Suspensions

The steady shear behavior of as-received Si_3N_4 suspensions at various solid
volume fractions is shown in Fig. 2. All of the suspensions are characterized by a
shear thinning which is associated with the disruption of suspension structure or
shear induced alignment. At lower shear rates, the suspension structure is close to
equilibrium because thermal motion dominates over the viscous force. At higher
shear rates, the viscous force affects the suspension structure and shear induced
alignment develops, and thus shear thinning occurs. On the other hand, for low
solid loading suspension, the viscosity curve develops a constant viscosity at
higher shear rate that commonly defined high-Newtonian region, which is the
indicative of completely hydrodynamically controlled suspension structure. With
increasing solid loading, high Newtonian region disappears gradually and shear
thickening appears instead at high shear rate. The shear thickening maybe results
from a flow instability which causes the particles to break out of their ordered
layers at some critical level of shear stress (i.e. shear rate), thereby causing the
rise in viscosity [18].

Attempts have been made to apply rheological models to the suspensions.
The experimental results indicate that any model shows a good fit to the
rheological behavior of suspensions only within a specific shear rate range.
Among the various rheological models existing in literature, including Bingham
model, Herschel-Buckley model, Sisco model, and Casson model, the Sisco

Table I Fitting parameters to Sisco model for as-received Si_3N_4 suspensions at various solid loading

ϕ	$\eta_\infty/Pa \cdot s$	$k/Pa \cdot s^n$	n
0.10	0.0013	0.072	0.92
0.15	0.0019	0.53	0.83
0.20	0.0032	1.3	0.65
0.25	0.0073	2.9	0.27
0.30	0.017	5.9	0.13

Table II Fitting parameters to Casson model for as-received Si_3N_4 suspensions at various solid loading

ϕ	τ_y/Pa	$\eta_c/Pa \cdot s$
0.35	2.5	2.6
0.40	3.4	6.1
0.45	9.4	53

Table III Fitting parameters to Sisco model for modified Si_3N_4 suspensions at various solid loading

ϕ	$\eta_\infty/Pa \cdot s$	$k/Pa \cdot s^n$	n
0.30	0.069	1.7	0.76
0.35	0.025	2.3	0.39
0.40	0.078	9.8	0.15

Table □ Fitting parameters to Casson model for modified Si_3N_4 suspensions at various solid loading

ϕ	τ_y/Pa	$\eta_c/Pa \cdot s$
0.45	7.2	1.7
0.50	12.8	2.8
0.52	35.3	8.6
0.54	150	11.6

model provides the best fit in the range of ϕ 0.30 while the Casson model in the range of 0.35 ϕ 0.45 within a shear rate range of 0.001~800 s^{-1} and the fitting parameters of these two models are given in Table I and II, respectively. Sisco equation is $\eta = \eta_\infty + k \cdot \gamma^{n-1}$. At $n=0$, the equation transforms into $\eta = \eta_\infty + k \cdot \gamma^{-1}$, i.e., $\tau = k + \eta_\infty \cdot \gamma$, which is just Bingham model. It is evident that parameter k and η_∞ represent yield stress and limiting viscosity at high shear rates respectively. At $n=1$, the equation transforms into $\eta = \eta_\infty + k$, which is Newtonian equation exactly. The increase of parameter k from 0.072 to 5.9 and the decrease of parameter n from 0.92 to 0.13 indicate that the yield stress increases gradually and the rheological behavior transforms from near Newtonian fluid into plastic fluid with increasing solid loading from 10 to 30vol%.

Casson equation is $\tau^{1/2} = \tau_y^{1/2} + (\eta_c \cdot \gamma)^{1/2}$, where parameter τ_y and η_c represent yield stress and Casson viscosity respectively. The yield stress τ_y can be used as a parameter that quantifies the strength of suspension structure. As is expected, the yield stress τ_y and the Casson viscosity η_c increase simultaneously with increasing solid loading of suspensions.

The steady shear behavior of modified suspensions is very similar to that of as-received suspensions, as is illustrated in Fig. 3. The shear behavior of modified suspensions fits to Sisco model in the range of ϕ 0.40 and Casson model in the range of 0.45 ϕ 0.54 perfectly. The fitting parameters of these two models are given in Table III and □, respectively. Comparison to that of as-received suspensions, the rheological behavior of modified suspensions is improved efficiently. For certain solid loading, the viscosity of modified suspension at any shear rate is lower than that of as-received suspension. Additionally, the 40 vol% as-received suspension exhibits shear thickening at high shear rate while the modified suspension develops still a high-Newtonian region even with a solid concentration of 45%. These effects of surface modification on the rheological behavior of Si_3N_4 suspension can be interpreted by taking the following three factors into account. Firstly, the zeta potential of Si_3N_4 particles in the alkaline region increases evidently in magnitude (Fig. 1), so the fluidity of suspension improves efficiently. Secondly, it is well known, there exist some silanol groups on the surface of Si_3N_4 particles that usually form an immobile water layer on the particle surface by hydrogen bond with water molecular [19]. As-received powder has more silanol group at the particle surface, which could result in multi-molecular adsorbed layer, increased efficient solid volume fraction and bad fluidity. On the other hand, modified powder has less silanol groups at the particle surface, which could result in mono-molecular adsorbed layer and improved fluidity. Additionally, the ICP results show that the amount of ferric oxide in modified powder decreased by 47% over the as-received powder. Thus, the decrease of the high valence counter-ion concentration (e.g., Fe^{3+}) is also responsible for improving the rheological behavior of suspension to a certain extent.

3.3 Viscoelastic Behavior of Si_3N_4 Suspensions
Viscoelastic response often presents for concentrated suspensions because of the interaction between suspended particles [10]. Viscoelastic measurements can be performed to characterize the rheological behavior, in particular the effect of solid volume fraction. Oscillatory viscoelastic measurements are meaningful only under small amplitude of oscillation, so that the suspension structure is only slightly perturbed from rest. Fig. 4 illustrates the results of a typical strain sweep experiment of as-received Si_3N_4 suspension with a volume fraction of 40 vol%, showing that the behavior of the suspension changes from an elastic response (G' remains constant) to a flow response (G' decreases) with increasing strain amplitude. The elastic modulus G' is a measure of elastically stored energy through particle-particle interaction while viscous modulus G'' quantifies dissipated energy through particle-medium or medium-medium interactions. The viscous modulus G'' is attributed to bulk flow. Fig. 4 shows the linear viscoelastic regime below the critical strain, γ_c, where the elastic modulus G' starts to drop sharply. The effect of strain amplitude on the elastic modulus G' of the as-received

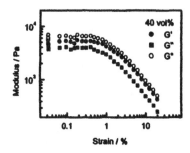

Fig. 4 Storage G', loss G" and complex G* as a function of the strain amplitude for 40vol.% Si₃N₄ slurries in the dynamic tests

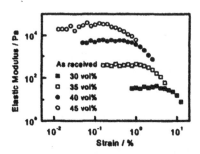

Fig. 5 Storage G' as a function of the strain amplitude for as-received Si₃N₄ suspensions with various solid loading in the dynamic tests.

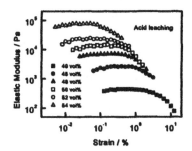

Fig. 6 Storage G' as a function of the strain amplitude for modified Si₃N₄ suspensions with various solid loading in the dynamic tests.

Si₃N₄ suspensions and that of the modified Si₃N₄ suspensions at various solid volume fraction are given in Fig. 5 and Fig. 6, respectively. As can be seen, the critical strain γ_c decreases continuously with the solid volume fraction. A similar behavior was found in a previous study on sterically stabilized suspensions of high concentration [20]. For certain solid volume fraction, the linear viscoelastic region of the modified suspensions extends and the corresponding G' decreases sharply in comparison with that of the as-received suspensions. All the following frequency sweep measurements are performed at low strain amplitudes, $\gamma < \gamma_c$, where the viscoelastic response is linear.

Fig. 7 shows the effect of solid volume fraction of the as-received suspensions on the frequency dependence of the elastic modulus G' and the viscous modulus G". In the frequency range examined, both moduli increase with frequency and the magnitude of the viscoelastic response increases with solid volume fraction. The viscoelastic response of the suspensions progressively changes from more viscous to more elastic. Similar behavior has been observed in previous studies [20-21]. For the lowest concentration, ϕ=0.35, the suspension is

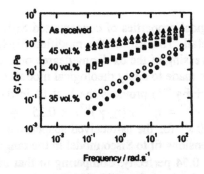
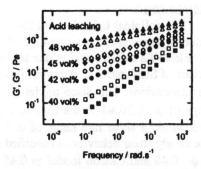

Fig. 7 Storage G' and loss G" (filled and unfilled symbols, respectively) as a function of the frequency for as-received Si_3N_4 suspensions

Fig. 8 Storage G' and loss G" (filled and unfilled symbols, respectively) as a function of the frequency for modified Si_3N_4 suspensions

essentially viscous in nature (G'<G") over the entire frequency examined. With increasing concentration to ϕ=0.40, the curve exhibits a characteristic frequency, ω_c, where G'=G". With increasing concentration, the crossover shifts to lower frequencies. For the highest concentration, ϕ=0.45, the elastic modulus G' exceeds the viscous one G" over the entire frequency range (G'>G"). The frequency sweep of the modified suspensions exhibits a very similar behavior to that of the as-received ones, as illustrated in Fig. 8. Comparison to that of the as-received suspensions, the modified suspensions transmit from viscous to elastic at higher concentration and the characteristic frequency ω_c shifts to higher frequency for certain solid concentration.

On the other hand, the inverse of the characteristic frequency ($1/\omega_c$) provides a measure of the structural relaxation time of the suspension, t_r [20]. Since a simple mechanical model like the Maxwell model cannot depict the viscoelastic response of concentrated suspensions exactly, the viscoelastic measurements only yield an average t_r. It is well known that the structural relaxation time t_r increases with solid concentration of suspensions. At lower frequency, the characteristic experimental time (t_o) of the suspension longer than its t_r ($t_o > t_r$), thereby the perturbed structure formed during dynamic oscillatory measurement relaxes well. Thus, the suspension dissipates the most of energy and exhibits smaller elastic modulus. With increasing oscillatory frequency, the characteristic experimental time t_o shortens than its t_r ($t_o < t_r$), the perturbed structure cannot be relaxed perfectly. The energy is stored in the suspension so it exhibits more elastic.

CONCLUSIONS

The colloidal behavior and the rheological properties of concentrated Si_3N_4 aqueous suspensions are investigated in detail. The IEP of the modified Si_3N_4 particle shifts to basic region due to the decreasing of the oxidizing degree on the surface of Si_3N_4 particle. Attempts have been made to apply rheological models to the suspensions. The Sisco model ($\eta = \eta_\infty + k \cdot \gamma^{n-1}$) provides the best fit in the range of ϕ 0.30 while the Casson model ($\tau^{1/2} = \tau_y^{1/2} + (\eta_c \cdot \gamma)^{1/2}$) in 0.35 ϕ 0.45 within a shear rate range of $0.001{\sim}800$ s^{-1} for the as-received suspensions. The steady shear behavior of modified suspensions fit to Sisco model in the range of ϕ 0.40 and Casson model in 0.45 ϕ 0.54 perfectly. Comparing to that of as-received suspensions, the rheological behavior of modified suspensions is improved efficiently. The dynamic oscillatory experimental indicates that the critical strain decreases, the linear viscoelastic regime narrows and the corresponding elastic storage modulus G′ increases continuously with increasing solid concentration. For the modified suspensions, the linear viscoelastic regime broadens and the corresponding G′ decreases sharply. With increasing solid concentration, the magnitude of the viscoelastic response increases, the characteristic frequency ω_c shifts toward lower frequencies and the suspension transforms from more viscous into more elastic. The transformation takes place at higher solid concentrations for modified suspensions. Moreover, the frequency dependence of the viscoelastic properties of suspensions can be interpreted by the concept of the structural relaxation time.

REFERENCES

1. F.F.Lang, "Powder Processing Science and Technology for Increased Reliability," *J. Am. Ceram. Soc.*, 72[1]3-15(1989)

2. C.P.Cameron and R.Raj, "Better Sintering Through Green-State Deformation Processing," *J. Am. Ceram. Soc.*, 73[7]2032-37(1990).

3. V.A.Hackley, "Colloidal Processing of Silicon Nitride with Poly(acrylic Acid): □, Adsorption and Electrostatic Interactions, □, Rheological Properties," *J. Am. Ceram. Soc.*, 80[9] 2315-25(1997), 81[9]2421-28(1998).

4. R.J.Pugh and L.Bergstrom (Eds.), *Surface and Colloid Chemistry in Advanced Ceramics Processing*. Marcel Dekker, New York, 1994.

5. P.C.Heimenz (Eds.), *Principles of Colloid and Surface Chemistry*. Marcel Dekker, New York, 1986.

6. M.P.Albano and B.Garrido, "Processing of Concentrated Aqueous Silicon Nitride Slips by Slip Casting," *J. Am. Ceram. Soc.*, 81[4]837-44(1998).

7. J.C.Chang, B.V.Velamakanni, F.F.Lange, and D.S.Pearson, "Centrifugal Consolidation of Al_2O_3 and Al_2O_3/ZrO_2 Composites Slurries vs Interparticle Potentials: Particle Packing and Mass Segregation," *J. Am. Ceram. Soc.*, 74[9]

2201-204 (1991).

8. L.Bergstrom, C.H.Schilling, and I.A.Aksay, "Consolidation Behavior of Flocculated Alumina Suspensions," *J Am Ceram Soc*, 75[12]3305-14(1992).

9. B.V.Velamakanni and F.F.Lange, "Effect of Interparticle Potential and Sedimentation on Particle Packing Density of Bimodal Particle Distributions During Pressing Filtration," *J. Am. Ceram. Soc.*, 74[1]166-72(1991).

10. L.Bergstrom, "Rheology of Concentrated Suspensions"; pp. 193-244 in *Surface and Colloid Chemistry in Advanced Ceramics Processing*. Edited by R.J.Pugh and L.Bergstrom, Marcel Dekker, New York, 1994..

11. W.B.Russel, D.A.Saville, and W.R.Schowalter, *Colloidal Dispersions*. Cambridge University Press, Cambridge, U.K., 1989.

12. R.Darby, "Hydrodynamics of Slurries and Suspension"; pp. 49-65 in *Encyclopedia of Fluid Mechanics*. Edited by N.P.Cheremisinoff, Gulf Publishing, Houston, TX, 1986.

13. X.J.Liu, L.P.Huang, X.Xu, H.C.Gu, and X.R.Fu, "Effects of Acid Leaching on Rheological Behavior of Silicon Nitride Aqueous Suspension, *J. Mater. Sci. Letter*, 19(3) 177-78 (2000).

14. X.J.Liu, L.P.Huang, X.R.Fu, and H.C.Gu, "Effects of Dispersant on Rheological Behavior of Silicon Nitride Aqueous Suspension, *J. Mater. Sci. Letter*, 18, 841-42(1999).

15. L.Bergstrom and E.Bostedt, "Surface Chemistry of Silicon Nitride Powders: Electrokinetic Behavior and ESCA Studies," *Colloids Surf*, 49, 183-97(1990).

16. X.J.Liu, L.P.Huang, X.W.Sun, X.Xu, X.R.Fu and H.C.Gu, "Rheological Properties of Aqueous Silicon Nitride Suspension" *J. Mater. Sci*. 36, 3379-84 (2001).

17. K.I.Palph (Eds.), *The Chemistry of Silica*. John Wiley & Sons, New York, 1979, pp. 56 & 195.

18. R.L.Hoffman, "Discontinuous and Dilatant Viscosity Behavior in Concentrated Suspensions. I. Observations of a Flow Instability," *Trans. Soc. Rheol.*, 16, 155-73(1972).

19. R.K.Iler and R.L.Dalton, "Degree of Hydration of Particles of Colloidal Silica in Aqueous Solution," *J Phys. Chem.*, 60, 955-57(1956).

20. L.Bergstrom, "Rheological Properties of Concentrated, Nonaqueous Silicon Nitride Suspension," *J Am. Ceram. Soc.*, 79[12]3033-40(1996).

21. D.A.R. Jones, B.Leary, and D.V.Boger, "The Rheology of a Concentrated Colloidal Suspension of Hard Spheres," *J. Colloid Interface Sci.*, 147, 479-95 (1991).

Si₃N₄ POWDERS APPLIED FOR WATER-BASED DCT

Si$_3$N$_4$ POWDERS APPLIED FOR WATER-BASED DCT

Ola Lyckfeldt
Swedish Ceramic Institute
P.O. Box 5403
SE-402 29 Göteborg
Sweden

Kent Rundgren
Permascand AB
P.O. Box 42
SE-840 10 Ljungaverk
Sweden

ABSTRACT

Two Si$_3$N$_4$ powders, one direct-nitrided, (P95L, Permascand AB) and one imide derived, high grade (SN-E10, UBE Industries, Japan) have been processed and evaluated to meet the demands in water-based Direct Casting Techniques (DCT).

Pre-dispersing followed by freeze granulation/freeze drying of SN-E10 was shown to significantly improve the possibilities of reaching high solids loading without critical slip dilatancy. Generally, pure electrostatic stabilization by a pH adjustment to 10, *in-situ* with P95L or with NH$_4$OH, was shown to be the more efficient dispersing concept. The use of polyacrylates gave considerably higher viscosity but a lower degree of dilatancy at extreme solids loading. Together with pre-milling/freeze granulation/freeze drying a solids loading of 54 vol% P95L (including 6 wt% Y$_2$O$_3$ and 2 wt% Al$_2$O$_3$ as sintering aids) could be reached using small amounts of polyelectrolyte.

Applied in DCT, e.g. protein forming (PF) and starch consolidation (SC), the addition of protein or starch resulted in increased viscosity but reduced dilatancy. Combined with a polyelectrolyte, pre-agglomeration of the protein took place, which gave less good sintering of the shaped Si$_3$N$_4$ specimens owing to a coarser microstructure. For SC the polyelectrolyte appeared to improve the dispersing of the starch granules which favored the sintering performance. With a pH adjustment to 10, protein-formed specimens sintered to near full density with gas pressure sintering whereas starch-consolidated specimens showed residual porosity originating from the starch granules.

INTRODUCTION

Silicon nitride (Si$_3$N$_4$) is today a structural ceramic that has an established market in engineering applications[1-2]. Several producers of Si$_3$N$_4$ powders and

Si_3N_4 components, including RBSN (reaction bonded Si_3N_4) with silicon powder as raw material, exist worldwide. The prospective of Si_3N_4 is also considered to be favorable with a growing market, not only for structural applications but also as functional materials. The manufacture of Si_3N_4 components requires careful control of each processing step from the raw powder properties, colloidal processing, shaping to sintering to achieve the required material properties. The raw powder properties in terms of purity, phase composition, surface silica (surface-chemical properties) and particle size distribution very much depend upon the specific manufacturing route (imide decomposition, carbothermal reduction, direct nitridation etc)[3-4]. The choice of powder is a matter for consideration in terms of cost, processing route (type of shaping and sintering) and material requirements. As there has been a growing pressure on cost reduction in the manufacture of Si_3N_4 components attention has focused on cheaper (direct-nitrided) powders, water processing and low pressure sintering. This often requires pre-processing to obtain a powder that is easy to disperse to high solids loaded suspensions at the same time as it gives adequate sintering performance.

Several research groups have studied dispersing of various silicon nitride powders in water over the last 15–20 years[5-7]. It is typical that commercial Si_3N_4 powders consist of various amounts of silica on the particle surfaces, which has a significant impact on the surface-chemical properties. The degree of surface silica influences the charging behavior at a specific pH, the isoelectric point (iep) and the degree of adsorption of dispersants. Depending on the powder-manufacturing route the formation of silica can bind particles into tight agglomerates[8]. These agglomerates entrap water, immobilize milling media, and can make it difficult to reach high solids loaded suspensions. One way to overcome this problem is to pre-disperse (de-agglomerate) at moderate solids loading, freeze and freeze dry to avoid re-agglomeration[9]. When re-dispersing, the powder shows less water binding capacity through the absence of tight agglomerates. Combined with an efficient stabilizing concept (dispersant or pH adjustment) higher solids loadings can therefore be reached.

A complicating factor when processing Si_3N_4 materials is the need of substantial amounts of sintering aids, often various oxides. Typical is yttria (Y_2O_3) that cannot be processed in water unless high pH is applied owing to a critical solubility that causes slip destabilization. Several successful attempts to inhibit solubility of yttria and silica on the Si_3N_4 particles by surface modifications have been made. One example is the use of silanes that has been shown to reduce the solubility significantly[10-11]. However, the processing appears complicated and inconvenient, especially when very high particle concentrations are to be reached.

Efficiently de-agglomerated and highly concentrated powder suspensions are favorable in most ceramic processing. This is crucial not least when using the so-called Direct Casting Techniques (DCT). In DCT (gel casting, direct coagulation casting (DCC), hydrolysis-assisted solidification (HAS) etc.)[12] shaping is conducted in non-porous molds by transforming a ceramic powder suspension into rigidity without compaction or removal of liquid. Consequently, the solids loading of a prepared suspension corresponds directly to the shaped density. One advantage of DCTs over other shaping techniques is the fact that consolidation takes place in a well-dispersed and homogeneous state. This promotes sintering, symmetric shrinkage and the ultimate material properties. Furthermore, components with complicated shapes and sections with varied thickness are possible to manufacture. Today, there are DCTs available that potentially will make it possible to combine non-hazardous and environmentally friendly processing with economic efficiency and high material demands. Protein forming (PF)[13] and starch consolidation (SC)[14] are two examples of water-based DCTs that utilize non-hazardous processing aids, such as globular proteins and starch. In PF the rigidity of a ceramic suspension relies on the gel formation of a globular protein (albumin, whey proteins etc) whereas the water absorption and the swelling of starch are responsible for the consolidation in SC. In PF, the nanosized, spherically configured protein molecules and a fine-stranded gel network normally provide full densification of shaped ceramics using low pressure sintering. SC, on the other hand, requires hot (isostatic) pressing in order to reach complete densification owing to the involvement of large starch granules (5–100 μm) that leave pores of the corresponding size after debinding.

In this study two Si_3N_4 powders, a medium-cost, direct-nitrided powder (SicoNide P95 L, Permascand AB) and a high-grade, imide-derived powder (SN-E10, Ube Industries Ltd), were evaluated and processed to meet the demands in DCT. The powders were used both in the as-received and in the pre-treated states for dispersing and forming studies. Utilizing DCT, exemplified by protein forming (PF) and starch consolidation (SC), material specimens were shaped and pressureless sintered or gas pressure sintered (GPS).

MATERIALS AND EXPERIMENTALS
Materials and Powder Pre-Treatments

The manufacturers' data on the Si_3N_4 powders used in this study are presented in Table I. Pre-treatment of P95L was conducted by milling for 48 h in water at 40 vol% without dispersant followed by freezing and freeze drying (Lyovac GT2, Leybold AB, Sweden). The freezing was carried out by spraying the suspension into liquid nitrogen, i.e. freeze granulation (LS-2, PowderPro HB, Sweden). Pre-treatment of SN-E10 was conducted by dispersing in water at 35 vol% with a pH adjustment to 10 with NH_4OH and subsequently freeze-granulating and freeze-drying. Dispersing was conducted by planetary milling (PM 400, Retsch,

Germany), using Si_3N_4 liners and balls, at 200 rpm for 60 min whereas 100 rpm was used for the milling.

Table I. Specifications for as-received SicoNide P95L and UBE SN-E10

Powder	$\alpha/(\alpha+\beta)$	Si (wt%)	Fe (wt%)	Al (wt%)	Ca (wt%)	C (wt%)	O (wt%)	BET area (m^2/g)
P95L	93/7	0.2	0.04	0.06	0.01	0.30	0.46	6.3
SN-E10	99.5	-	<0.01	<0.005	<0.005	-	1.33	11.0

Other ceramic powders used as sintering additives were Al_2O_3 (AKP-30, Sumitomo Corp., Japan) and Y_2O_3 (Grade C, HC Starck GmbH, Germany).

Dispersants (polyelectrolytes) or pH adjustment to 10 with NH_4OH were used as dispersing concepts. The utilized polyelectrolytes, Duramax D-3021 from Rohm & Haas and Dolapix PC75 from Zschimmer & Schwarz, are assumed to be polyacrylates, a type of dispersant commonly used for water-based processing of Si_3N_4. In the case of P95L, the natural pH adjustment to about 10 by the powder itself made it unnecessary to use NH_4OH.

As consolidators, a globular protein (Bovine Serum Albumin (BSA), A5401, Sigma Chemicals Ltd) and a potato starch (Mikrolys 54, Lyckebystärkelsen AB, Sweden) were utilized.

Powder Characterizations

Zeta (ζ) potential measurements in the range of pH 3–11 with as-received and pre-treated Si_3N_4 powders were done with an Acoustosizer (Matec Applied Sciences). For this, 5 vol% suspensions were prepared with a background electrolyte concentration of 0.01 M KCl.

The effect of the milling of P95L upon the physical properties was characterized by measurements of BET specific surface area (FlowSorb II 2300, Micromeritics, USA) and particle size distributions (Sedigraph 5100, Micromeritics, USA).

Dispersing and Consolidation Experiments

Powder suspensions at various solids loadings of as-received and milled/ freeze-granulated/freeze-dried Si_3N_4 powders were prepared by planetary milling at 200 rpm for 60–120 min. Higher solids loading was reached by adding the powder in portions which required the longer total milling time. In the case of P95L, 6 wt% Y_2O_3 and 2 wt% Y_2O_3 as sintering aids were included in all suspensions in order to evaluate a realistic powder composition. After milling, all slips were conditioned for 16 h in slowly rotating plastic containers without balls before rheological evaluations took place.

Either 10 wt% protein (BSA) or 3 vol% starch based on water was added to slips with higher solids loadings of P95L. The slips were impeller stirred for 2 h prior to rheological evaluations.

Rheological Measurements

Rheological studies (viscosity and viscoelasticity) were carried out with a rotational controlled-stress rheometer (StressTech, ReoLogical Instruments AB, Sweden) using a concentric cylinder measurement device (Ø = 25 mm) with a 1 mm gap. Steady-shear measurements (equilibrium viscosity) of the suspensions were conducted in the shear rate range of 1–700 s^{-1}. To achieve equal rheological history, all suspensions were exposed to a pre-shearing at 400 s^{-1} for 1 min, followed by a rest for 1 min prior to measurement. The characterization of the consolidation, the protein gelling and the starch-swelling processes, was carried out by oscillatory shear measurements at a constant frequency (1 Hz) and strain (10^{-3}) while the temperature was increased by 2°C/min from 25°C to 70°C or 80°C (15 min dwell time).

Shaping and Sintering

A few drops of defoamer (Contraspum, Zschimmer & Schwarz GmbH) were added to the P95L slips with protein or starch, and the slips were vacuum-treated (at about 100 Pa) for a short period of time until, visually determined, the air bubbles were removed. Shaping of specimens (Ø = 10–25 mm, h = 10 mm) was carried out in covered plastic rings placed on a Teflon surface. To achieve consolidation, the molds were treated at 70°C (SC) or 80°C (PF) for 60 min. After cooling to ambient temperature demolding took place and drying at ambient conditions was conducted. Removal of the remaining water and organic additives (starch or protein) was carried out either in a nitrogen atmosphere or in air with 1°C/min up to 500°C and 60 min dwell. Using a graphite resistance furnace (Balzer/Pfeiffer, Germany) specimens were pressureless sintered, placed inside a Si_3N_4 powder bed in a graphite crucible, at 1800°C at 0.1 MPa nitrogen pressure for 3 h. Some specimens were gas pressure sintered (FPW 250/300, FCT, Germany) placed inside a Si_3N_4 powder bed in a Si_3N_4 crucible. In this case a two-step sintering cycle, 3 h and 1 MPa at 1800°C followed by 3 h and 3 MPa at 1900°C, was utilized. The densities of sintered specimens were measured using Archimedes' principle (the water intrusion method).

RESULTS and DISCUSSIONS

Effects of Powder Pre-Treatments

The results from the ζ potential measurements with the SN-E10 powder versions are presented in Figure 1. For the as-received powder the isoelectric point (pH$_{iep}$) was defined to about 5.4. Pre-dispersing of SN-E10 resulted in a slight increase of pH_{iep} to about 5.8 and a decrease of the ζ potential at high pH. This indicates that a certain removal (leaching) of surface silica occurred during the pre-dispersing procedure. It has to be noted that the dwell time for the powder in water was in fact longer (2–3 h) than the specific dispersing (milling) step prior to the freeze-granulation operation. The reduced number of silanol groups, which is responsible for the negative charge sites on the surfaces, is also expected to reduce the zeta potential at high pH levels.

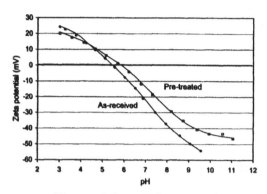

Figure 1. Measurement of ζ potential vs pH for as-received and pre-treated UBE SN-E10. Adapted from[9].

In contrast, the pH_{iep} decreased from about 5.3 to 4.8, and the charge at high pH increased for the P95L powder through the milling procedure (see Figure 2). In this case (new) Si_3N_4 surfaces were produced and, when exposed to hydrolysis reactions, an overall increase of the silica content could be expected. With small additions of the polyelectrolyte PC75 a clear shift of the pH_{iep} downwards can be seen. This is most likely a result of polymer adsorption at lower pH. However, at high pH the level of charging is similar to that without dispersant indicating that no or limited polymer adsorption took place. This is in agreement with the findings of others[15-16] regarding adsorption of polyacrylic acids on Si_3N_4.

Besides the change in surface charge properties the milling of P95L resulted in an increase of the BET surface area from 6.3 to 10.1 m^2/g and a reduction of the mean particle size (d_{50}) from 1.3 to 0.7 μm (Sedigraph 5100, Micromeritics, USA). The finer and narrower particle size distribution was expected to promote the sintering unless the indicated increase of silica unfavorably changed the sintering aid composition.

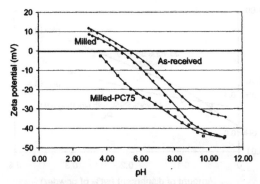

Figure 2. Measurement of ζ potential vs pH for as-received and pre-treated P95L, the latter with and without dispersant (0.05 wt% PC75).

Dispersing Study

Figure 3 shows the low-shear viscosity of Si_3N_4 slips based on as-received SN-E10 and P95L with various amounts of dispersant. The zero-point levels of polymer addition represent a pH adjustment to 10, occurring naturally with P95L and through the addition of a base with SN-E10. The difference in solids loadings should be noted along with the fact that the P95L suspensions included sintering aids. P95L caused a natural pH adjustment to >10 by a significant hydrolysis with a release of NH_3 that is far from being as pronounced as when using SN-E10. Furthermore, both dispersants adjust the pH to a similar level, also given in the figure.

In both cases the addition of the polyelectrolyte gave rise to a continuous viscosity increase, stronger for the SN-E10 than for the P95L system. Since the adsorption of polyelectrolytes, such as polyacrylic acids (PAA), on Si_3N_4 is poor at high pH levels, the influence of free and highly dissociated polymers must be the reason for these results.

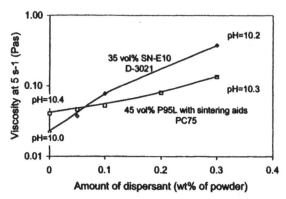

Figure 3. Steady shear viscosity at low shear rates vs amount of polyelectrolyte added to Si₃N₄ slips based on SN-E10 and P95L (with sintering aids).

The viscosity increase could be due to interactions between free polymer molecules and a general increase of the water-phase viscosity. However, the destabilizing phenomena caused by an increased electrolyte concentration and charge-screening effects are also likely to be significant. The negative consequences of adding polyelectrolyte appeared to be less serious with the P95L system. But in this case we could expect a certain polymer adsorption on the sintering aids (Al_2O_3 and Y_2O_3) that reduced the amount of free polymer and contributed to a stabilization of the oxide particles. In general, the P95L powder is a more easily processed powder that can be deagglomerated with less energy input than SN-E10. Together with a wider particle size distribution, higher solids loadings can therefore be reached with P95L, and this is reflected by the viscosity data in Figure 3. In DCT it is desirable to maximize the solids loading to achieve high shaped density. However, in view of the processing requirements, the viscosity level as such but more often a critical slip dilatancy is the limiting factor. Dilatancy is a common phenomenon with highly concentrated and well-stabilized powder suspensions. For the Si_3N_4 powders used in this study, dilatancy appeared at different levels of solids loading as illustrated in Figure 4 and 5. Regarding the as-received powders, electrostatically stabilized at high pH levels, dilatancy appeared at about 50 vol% P95L and about 42 vol% SN-E10. This situation was changed when the pre-treated powder versions were used, especially for SN-E10. Apart from a significant viscosity decrease, dilatancy with the pre-dispersed SN-E10 did not become pronounced until a solids loading of about 46 vol% was reached. Figure 4 shows that the viscosity profile still expresses a pure shear thinning behavior at 44 vol% solids.

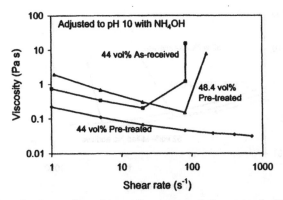

Figure 4. Steady shear viscosity vs shear rate of electrostatically stabilized Si_3N_4 slips based on UBE SN-E10. Adapted from[9].

Since the SN-E10 powder became easier to deagglomerate by the pre-treatment we could also expect a substantial difference in the dispersing process as the solids loading became high. In a situation with the same powder concentration, another addition of the tight agglomerates of as-received powder would rapidly entrap water and immobilize milling media. The consequence would be a significant increase of viscosity and dilatancy that would limit or even inhibit further deagglomeration. The dilatancy in itself would be a factor that immobilizes the milling media under high-speed conditions.

The impact of milling on the P95L slips became less pronounced (see Figure 5) than the pre-treatment of SN-E10 showed. Since the powder is easy to deagglomerate already in the as-received state the differences in rheological behavior are mainly related to the change in particle size distribution towards a finer and narrower size range. In practice, this will give an increase of the effective solids loading. The effective volume, including the electrostatic double-layer (the range of the electrostatic forces) of a smaller particle is always influenced more than that of a bigger one. Consequently, we could expect a general viscosity increase as is shown in Figure 5. More important for the processing properties, however, is the decrease in dilatancy by the milling operation. As indicated by the ζ potential measurements, an increased amount of surface silica also was assumed to result in an increased ion concentration. Like the solubility of common sintering aids such as Y_2O_3, soluble silica is an often-addressed ageing problem in slurry processing of Si_3N_4. However, at extreme solids loading an increased ion concentration through the presence of a larger amount of surface silica can contribute to decreased dilatancy by charge screening and a compressed electrostatic double-layer.

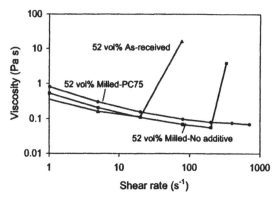

Figure 5. Steady shear viscosity vs shear rate of electrostatically stabilized Si_3N_4 slips based on P95L including sintering aids. Adapted from[17].

Since polyacrylates have shown poor adsorption onto Si_3N_4 at high pH, the main stabilizing factor must be of electrostatic origin through the pH-adjusting ability of the commonly used PAA. The reason for the results shown in Figure 5, where the addition of 0.05 wt% Dolapix PC75 gave a general viscosity increase but, at the same time, a clear decrease of the dilatancy, can therefore be related to ion-concentration effects. Rheological impact by free-polymer interactions is also possible but considered to be less important. At even higher solids loading (54 vol%) the rheological difference, with and without dispersant, was almost eliminated (see Figure 6). When approaching the maximum solids loading the ion concentration in the water phase will increase in general. A further contribution to the ion concentration by added polyelectrolyte would therefore have less impact.

Protein and Starch Additions

The addition of the globular protein (BSA) to highly concentrated Si_3N_4 suspensions pushes the viscosity further up as shown in Figure 6. This effect can be seen as a result of increased solids loading through the addition of spherically configured polymers. Depending on the specific system (dispersant, type and concentration of ions) the protein molecules have a tendency to agglomerate to a varying degree. In this case, the viscosity increase was more severe with the dispersant (PC75) present, which indicated an interaction between the free dispersant polymer and the protein molecules. However, as the viscosity increased, the degree of dilatancy decreased when adding BSA and even more in combination with the dispersant, which is a favorable result for the slip processing.

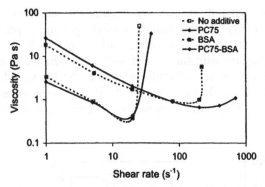

Figure 6. Steady shear viscosity for 54 vol% pre-milled P95L (including sintering aids) before and after addition of 10 wt% BSA based on water. Adapted from[17].

When adding starch, the viscosity increased significantly. The main reason for this is the fact that starch granules absorb a certain amount of water already at room temperature. As in the case of protein, starch additions tended to decrease the degree of dilatancy. This, however, was less pronounced with dispersant present. The lower viscosity with PC75 indicated that the polyelectrolyte might have contributed to a more effective dispersing of the starch granules even though their highly hydrophilic character normally make them easy to disperse in water without dispersant. Another effect can of course be that dispersant adsorption on the starch granules inhibited water absorption or suppressed any attraction to Si_3N_4 particles and, therefore, limited the viscosity increase.

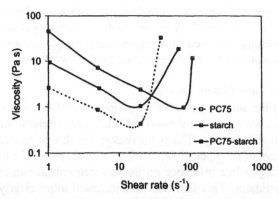

Figure 7. Steady shear viscosity for pre-milled P95L (including sintering aids) before and after addition of 3 vol% starch based on water. Adapted from[17].

As illustrated in Figure 8, sudden and significant increases of the storage modulus (dynamic rigidity) during heating indicate gelling of protein (BSA) or swelling of the starch. In general, the consolidation was much stronger and took place at lower temperatures with starch than with BSA. The water absorption and the swelling of starch were not critically influenced by the specific system, i.e. dispersants or ions present. On the other hand, BSA showed more sensitivity to the slurry composition. This was expressed by different gelling behavior with or without dispersant.

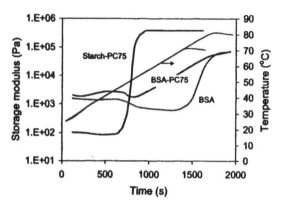

Figure 8. Storage modulus vs time and temperature for slips based on 54 vol% P95L (including sintering aids) with the addition of 3 vol% starch or 10 wt% BSA based on water. Adapted from[17].

With PC75 a prolonged gelling process took place, initiated at low temperatures by a pre-agglomeration. This might have resulted in a coarse gel structure. With no dispersant, the gelling occurred at a higher temperature but was more rapid which is assumed to result in a more fine-stranded gel network.

Evaluations of Shaped and Sintered Materials

The results from the sintering of selected shaped materials are summarized in Table II. The first obvious conclusion is that pressureless sintering was insufficient to reach full density. The main reason for this is related to the low shaped density, typical of all types of DCT. PF specimens showed a higher degree of densification owing to less influence on the microstructural homogeneity than with starch as consolidator. This fact was demonstrated more clearly when GPS was utilized, in which case PF materials reached near full density whereas the SC materials only reached 94% of theoretical density at best. To be able to sinter SC specimens of advanced ceramics to full density, hot isostatic pressing (HIP) is required. The main reason for this is the size of the starch granules (10–30 μm), which expand further by the water absorption. This results in large pores after debinding that remain after sintering. On the other hand, small protein molecules (some nm) can form networks within the water phase that normally keeps the homogeneity unaffected unless significant agglomeration occurs prior to gelling. However, higher initial slip viscosity and a prolonged gelling process when the polyelectrolyte (PC75) was present indicated such agglomeration. In turn, this gives rise to a coarser gel structure, less favorable shaped microstructure and, as shown in Table II, lower sintered density when PC75 was used. In contrast, the SC specimens in which PC75 had been used showed significantly higher sintered density than those without the polyelectrolyte. Again, the viscosity data indicated better dispersing of the starch and/or reduced water adsorption, factors that promoted sintering of shaped materials. Microstructural studies might give further confirmation of these results.

The results in Table II also indicate a positive effect of debinding in air compared to debinding in nitrogen. This was most pronounced with the SC specimens. The main reason is believed to be a more efficient debinding rather than an influence on the oxygen content. However, more studies are required to define the influence of the debinding conditions on the carbon/oxygen content.

Table II. Results from pressureless sintering at 1800°C and GPS (in bracket) of PF and SC specimens based on 54 vol% milled P95L powder. Adapted from[17].

Dispersant	Shaping	Effective solids (vol%)	Debinding atmosphere	Sintered relative density (% of theoretical)
-	PF	51.4	N_2	88.0
			Air	90.2 (99.4)
PC75	PF	51.4	N_2	84.1
			Air	88.2 (98.4)
-	SC	53.2	N_2	78.4
			Air	79.4 (86.4)
PC75	SC	53.2	N_2	79.5
			Air	80.8 (94.0)

SUMMARY AND CONCLUSIONS

This study has shown that the processing of Si_3N_4 powders in water at extreme solids loadings can be supported by powder pre-treatments and basic knowledge of colloidal effects. A powder that is highly agglomerated and difficult to process, such as UBE E10, can be given more favorable properties by pre-agglomeration and freeze granulation. Higher solids loading without critical dilatancy can thus be reached, which is crucial when utilizing direct casting techniques (DCTs). With direct-nitrided powders, such as SicoNide P95L (Permascand AB), higher solids loading can generally be reached owing to wider particle size distributions and less severe agglomeration in the as-received state. Increase of ζ_{iep} indicated leaching and reduction of SiO_2 at short-term processing of SN-E10 whereas the opposite effect was found for long-term milling of P95L. Typical of the latter powder was also a significant hydrolysis with ammonia release that produced a self-dispersing effect through the adjustment of the pH to a high level. As in the case of adjusting the SN-E10 slips to pH 10 with NH_4OH, the pure electrostatic stabilization was shown to be the most efficient dispersing concept. In general high pH is also required to restrict solubility of commonly used sintering aids, such as Y_2O_3, that may cause flocculation of the Si_3N_4 slip.

Considering poor polymer adsorption at high pH of commonly used dispersants for Si_3N_4 such as polyacrylic acids (PAA), the major dispersing effect is the pH-regulating ability. In this study the addition of polyelectrolytes gave rise to higher viscosity but less slip dilatancy at extreme solids loading than pH adjustment to 10. The milling of P95L with the assumed increase of the SiO_2 content resulted in similar rheological effects. The main factor behind these effects was concluded to be an increase of the ion concentration by the polyelectrolyte or by soluble silica that reduced the electrostatic interactions between the Si_3N_4 particles. Although the stabilizing force was negatively affected, the decrease of dilatancy was considered to be more important for the processing performance in shaping operations. Small amounts of PAA are therefore suitable when highly concentrated Si_3N_4 is to be processed, as they also enable accurate dispersing of sintering aids.

In this study, other processing (consolidating) additives were used as well and they were shown to have significant rheological impact on the Si_3N_4 suspensions. Starch addition and even more protein (Bovine Serum Albumin) addition tended to increase the viscosity but decrease the degree of dilatancy. The gelling of the protein was negatively influenced by the polyelectrolyte present that caused pre-agglomeration, an assumed coarser gel-structure and less favorable sintering performance of shaped specimens. For starch consolidation, polyelectrolyte gave the opposite effect with assumed better dispersing of the starch granules and better sintering performance of shaped specimens. In general, starch consolidation showed less sensitivity to the additives present, higher consolidation force but less good sintering of shaped specimens than protein forming did. Based on milled P95L powder and sintering additives (6 wt% Y_2O_3 and 2 wt% Al_2O_3) protein-formed specimens were sintered to full density if gas pressure sintering was utilized.

ACKNOWLEDGEMENTS

The authors wish to thank Permascand AB for the financial and technical support and the co-workers at the Swedish Ceramic Institute for their contributions to the experimental parts of the work.

REFERENCES

[1]E. Belfield, "Non-Oxide Advanced Ceramics Widen Their Application," *Global Ceramic Review*, 10–11 (2000).

[2]F. L. Riley, "Silicon Nitride and Related Materials," *Journal of the American Ceramic Society*, 83 [2] 245–65 (2000).

[3]H. Lange, G. Wötting and G. Winter, "Silicon Nitride-From Powder Synthesis to Ceramic Materials," Angew. Chem. Int. Ed. Engl. 30 (1579–97 (1991).

[4]T. Yamada, "Synthesis and Characterization of Sinterable Silicon Nitride Powder"; pp 15–27 in *Silicon-Based Structural Ceramics*, Ceramic Transactions, Volume 42. Edited by B. W. Sheldon and S. C. Danforth, The American Ceramic Society, Westerville, Ohio, 1994.

[5]R. De Jong, "Dispersion of Silicon Nitride Powders in Aqueous Medium"; pp. 477–484 in *Ceramic Powder Science II, A*, Ceramic Transactions, Volume 1. Edited by G. L. Messing, E. R. Fuller jr. and H. Hausner, The American Ceramic Society, Westerville, Ohio, 1988.

[6]G. Subhas, G. Malghan and L. Lima, "Factors Affecting Interface Properties of Silicon Nitride Powders in Aqueous Environment"; pp. 403–412 in *Ceramic Powder Science III*, Ceramic Transactions, Volume 12. Edited by G. L. Messing, S. Hirano and H. Hausner, The American Ceramic Society, Westerville, Ohio, 1990.

[7]P. Greil, "Review: Colloidal Processing of Silicon Nitride Ceramics"; pp. 319–327 in *Proceedings of the Third International Symposium on Ceramic Materials and Components for Engines*. Edited by V. J. Tennery, The American Ceramic Society, Westerville, Ohio, 1989.

[8]E. Laarz, B.V. Zhmud and L. Bergström, "Dissolution and Deagglomeration of Silicon Nitride in Aqueous Medium," *Journal of the American Ceramic Society*, 83 [7] 2394–2400 (2000).

[9]O. Lyckfeldt, L. Palmqvist and F. Poeydemenge, "Dispersing Si_3N_4 at High Solids Loading – Applied to Protein Forming"; pp. 75–78 in *Euro Ceramics VII*, Key Engineering Materials, Vols. 206–213, Trans Tech Publications, Switzerland, 2001.

[10]M. A. Buchta and W.-H. Shih, Improved Aqueous Dispersion of Silicon Nitride with Aminosilanes," *Journal of the American Ceramic Society*, **79** [11] 2940–46 (1996).

[11]M. Colic, G. Franks, M. Fisher and F. Lange, "Chemisorption of Organofunctional Silanes on Silicon Nitride for Improved Aqueous Processing," *Journal of the American Ceramic Society*, **81** [8] 2157–63 (1998).

[12]W. M. Sigmund, N. S. Bell and L. Bergström, "Novel Powder-Processing Methods for Advanced Ceramic," *Journal of the American Ceramic Society*, **83** [7] 1557–74 (2000).

[13]O. Lyckfeldt, J. Brandt and S. Lesca, "Protein Forming – a Novel Shaping Technique for Ceramics," *Journal of the European Ceramic Society*, **20**, 2551–59 (2000).

[14]O. Lyckfeldt and J. M. F. Ferreira, "Processing of Porous Ceramics by Starch Consolidation", *Journal of the European Ceramic Society*, **18**, 131–40 (1998).

[15]V. A. Hackley, "Colloidal Processing of Silicon Nitride with Poly(acrylic acid): I, Adsorption and Electrostatic Interactions," *Journal of the American Ceramic Society*, **80** [9] 2315–25 (1997).

[16]E. Laarz and L. Bergström, "The Effect of Anionic Polyelectrolytes on the Properties of Aqueous Silicon Nitride Suspensions," *Journal of the European Ceramic Society*, **20**, 431–40 (2000).

[17]O. Lyckfeldt and K. Rundgren, "High Solids Loaded Si_3N_4 Suspensions for Water Based DCT", Presented at CIMTEC 2002, To be published.

SYNTHESIS OF Si_2N_2O CERAMICS FROM DESERT SAND

M. Radwan and Y. Miyamoto
Joining and Welding Research Institute
Osaka University
Osaka 567-0047 Japan

H. Kiyono and S. Shimada
Graduate School of Engineering
Hokkaido University
Sapporo 060-8628 Japan

ABSTRACT

Si_2N_2O ceramics have been synthesized from a mixture of desert sand and reclaimed silicon by using the nitriding combustion-synthesis method which is known as an energy saving process. A model for the synthesis reaction based on both thermodynamic considerations and measured thermograms is discussed. The properties of fabricated porous and dense compacts are also presented.

INTRODUCTION

Silicon oxynitride was first reported in the 1950s[1,2] but its chemical composition as Si_2N_2O and its orthorhombic crystal structure were confirmed by Brosset and Idrestedt[3] in 1964. It is one phase in the sialons system[4] lying between Si_3N_4 and SiO_2. It is built up of SiN_3-O tetrahedral units joined by sharing corners and the orthorhombic crystal structure consists of irregular but parallel Si-N sheets linked together by Si-O-Si bonds, see Fig. 1[4]. Trigg and Jack[5] in 1987 pointed out that the Si_2N_2O forms a narrow solid-solution range with alumina (up to 10 mol % alumina at 1800°C), named O′-sialon with a general composition $Si_{2-x}Al_xO_{1+x}N_{2-x}$ in which Si^{4+} N^{3-} are replaced by Al^{3+} O^{2-} without a change of the crystal structure.

It was clear since the early work of Washburn[6] in 1967 that silicon oxynitride exhibits superior oxidation resistance to that of silicon nitride and silicon carbide especially at high temperatures. It shows good chemical stability for molten siliceous slags and molten metals. From 1967 to 1980s, Washburn in Norton Company had developed porous and dense silicon oxynitride components that were used as refractories and heating elements.[7-11] In the recent two decades, near full-dense silicon oxynitride materials with comparable thermal expansion coefficient and mechanical properties to Si_3N_4 were developed by several research groups[12-19] for the high temperature engineering applications. In a novel application, silicon oxynitride with high surface area was reported as a solid base catalyst for some organic reactions.[20]

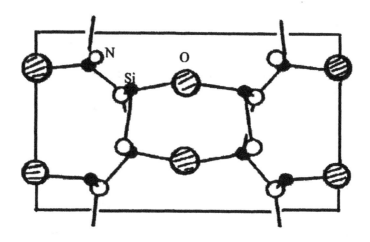

Figure 1. Crystal structure of Si_2N_2O

In spite of the above increasing interest, the thermodynamic properties of silicon oxynitride are not well known yet[21-24] and the material is not popular in use such as Si_3N_4. This can be attributed to the difficulties that still present in the current processing methods from both technical and economical views. Without doubt, there is no economic and reproducible method for the preparation of pure silicon oxynitride materials. The current methods for the preparation of silicon oxynitride are:

1- Silicothermal reduction of silica in presence of alkaline earth oxides at temperatures above 1450°C under controlled nitrogen-based atmosphere.

$$3/2 \, Si + ½ \, SiO_2 + N_2 \rightarrow Si_2N_2O \qquad (1)$$

This method was mainly developed in Norton Company and the product was usually impure having unreacted silicon of about 10 wt %.

2- Reaction sintering of an equimolar $Si_3N_4 + SiO_2$ mixture in the presence of liquid phase forming sintering aids at a temperature from 1600°C – 1950°C with or without using a pressure under controlled nitrogen gas.

$$½ \, Si_3N_4 + ½ \, SiO_2 \rightarrow Si_2N_2O \qquad (2)$$

High strength dense materials have been prepared, for example by Lewis[13], Mitomo[16], Ohashi[18], and Larker[19]. However, this method needs long heating times at high temperatures.

3- Carbothermal reduction of silica by carbon source at temperatures from 1400°C – 2000°C under controlled nitrogen-based atmosphere as was described by Trigg[25]:

$$2 \ SiO_2 + 3 \ C + N_2 \rightarrow Si_2N_2O + 3 \ CO \qquad (3)$$

This reduction reaction is usually incomplete and results in impure product containing unreacted phases.

4- Heating silica in controlled ammonia-based atmosphere[26] at temperature about 1300°C:

$$2 \ SiO_2 + 2 \ NH_3 \rightarrow Si_2N_2O + 3 \ H_2O \qquad (4)$$

In conclusion, there is still a necessity for an alternative method that can overcome these difficulties and produce as much pure as possible silicon oxynitride.

We found a new promising method that can be used for preparing near single-phase silicon oxynitride powder from a cheap starting mixture. It depends on the nitriding combustion-synthesis of silicon oxynitride from reclaimed silicon and desert sand mixture under pressurized nitrogen gas. The main advantages of this method are:

- The combustion-synthesis method, also known as the self-propagating high-temperature synthesis method or SHS, is an advanced process used successfully for the preparation of a variety of refractories and ceramics such as borides, nitrides, carbides, and silicides.[27,28] It depends on the strong exothermic heat release from the chemical reaction which propagates spontaneously and rapidly through the reactants converting them into a product. Simple equipment, high synthesis rate, and high-energy efficiency are its main characteristics.

- Desert white sand is a unique natural ore which is widely abundant, pure and cheap in the earth's crust.

In the following sections, the nitriding combustion-synthesis of Si_2N_2O from a mixture of reclaimed Si, desert sand and Si_2N_2O (as a diluent) will be reported. The main reactions involved in this system will be discussed based on the thermodynamic calculations. A brief report for the main properties of fabricated porous and dense Si_2N_2O compacts by using the pressureless sintering (PLS) and the spark plasma sintering (SPS) techniques, respectively, will be presented.

NITRIDING COMBUSTION SYNTHESIS OF Si₂N₂O POWDER

Experimental Procedure

The silica source used in this work was natural desert sand obtained from Sinai Peninsula of Egypt. It has high purity of > 99 % SiO_2 and was pulverized to – 40 μm size. The silicon used is a byproduct of zinc smelting industry from Toho Zinc Co. Ltd., Japan. Its particle size and purity are 8 μm and 94 %, respectively. Presynthesized Si_2N_2O powder having 5 μm size was used as a diluent. The mixing ratios of the reactants were calculated based on the following chemical reaction:

$$3(1-X)/2 \text{ Si} + (1-X)/2 \text{ SiO}_2 + X \text{ Si}_2\text{N}_2\text{O} + (1-X) \text{ N}_2 \rightarrow \text{Si}_2\text{N}_2\text{O} \tag{5}$$

Nitriding combustion experiments with different Si_2N_2O additions were carried out under 3 MPa nitrogen gas in a combustion apparatus (Fig. 2). Fifty grams of the reaction mixture was placed in a porous carbon crucible and the combustion reaction was initiated at the bottom end from the ignition of 5g thermite powder placed beneath the reaction mixture by passing 60 A current, for seconds, through carbon ribbon heater. The temperature profile and the combustion temperature, T_{max}, of the nitriding combustion reactions were measured by using W/Re thermocouple inserted in the reaction mixture.

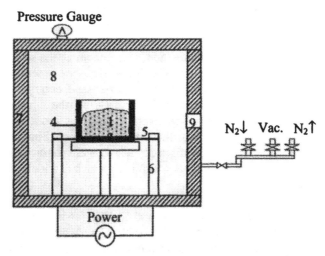

Figure 2. Schematic diagram of SHS apparatus (combustion furnace): (1) burden, (2) ignition agent, (3) porous container, (4) thermocouple, (5) carbon ribbon heater, (6) electrodes, (7) high-pressure chamber, (8) nitrogen gas atmosphere, and (9) glass window for observation

Product Phases and Morphology

The product of the combustion reactions consists of gray and hard roll surrounded by dark gray loose laminates. Without Si_2N_2O addition to the starting mixture (X = 0), the product had visible small melted residual silicon in the center. With Si_2N_2O additions (4, 6, and 8 wt %), the residual silicon decreases. At 10 wt % Si_2N_2O addition, no residual silicon was observed. The XRD patterns of the product in either case showed strong peaks for silicon oxynitride (JCPDS-file 47-1627) with minor or negligible peaks corresponds to β-Si_3N_4 (Fig. 3).

This set of experiments indicated that a mixture of reclaimed silicon and desert sand can sustain self-propagation combustion reaction under 3 MPa nitrogen gas forming near single phase silicon oxynitride powder and the addition of 10 wt % Si_2N_2O to the starting reaction promotes the formation of homogeneous Si_2N_2O. The addition of the product to the starting mixture is a useful method to reduce the exothermic heat and control the combustion reaction powders. Fig. 4 shows the scanning electron micrograph of the synthesized silicon oxynitride powder. The powder is agglomerates of fine particles with sizes < 5μm.

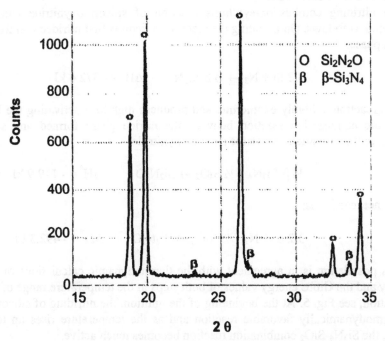

Figure 3. X-ray diffraction pattern of synthesized Si_2N_2O powder

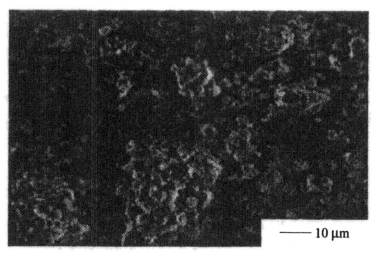

───── 10 μm

Figure 4. SEM micrograph of synthesized Si_2N_2O powder

Thermodynamic Considerations

The nitriding combustion-synthesis reaction of silicon oxynitride can be understood as follows. On initiating the reaction, silicon is first nitrided and forms β-Si_3N_4 phase.

$$3/2 \, Si + N_2 \rightarrow \tfrac{1}{2} \, \beta\text{-}Si_3N_4 \qquad \Delta H^\circ = -372.4 \text{ kJ} \qquad (6)$$

This reaction is highly exothermic and produces high heat activating the low exothermic combination reaction between the nitride phase formed and sand resulting in the formation of a stable silicon oxynitride.

$$\tfrac{1}{2} \, \beta\text{-}Si_3N_4 + \tfrac{1}{2} \, SiO_2 \rightarrow Si_2N_2O \qquad \Delta H^\circ = -119.9 \text{ kJ} \quad (7)$$

The net reaction is:

$$3/2 \, Si + N_2 + \tfrac{1}{2} \, SiO_2 \rightarrow Si_2N_2O \qquad \Delta H^\circ = -492.3 \text{ kJ} \quad (8)$$

This postulation is in agreement with the thermodynamic calculations of the enthalpy and the Gibbs energy values of both steps in the temperature range of the combustion, see Fig. 5. At the beginning of the ignition, the nitriding of silicon is the thermodynamically favorable reaction and as the temperature rises up to > 2000 K, the Si_3N_4-SiO_2 combination reaction becomes much active.

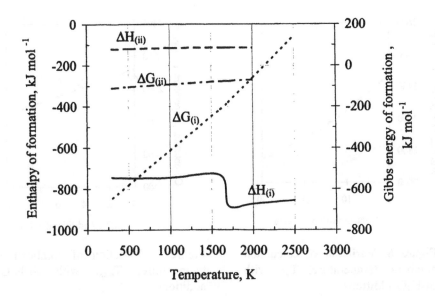

Figure 5. Enthalpy and Gibbs energy of formation of the reactions:
(i) $3Si + 2N_2 = Si_3N_4$, and (ii) $\frac{1}{2}Si_3N_4 + \frac{1}{2}SiO_2 = Si_2N_2O$

The calculated adiabatic temperature, T_{ad}, of silicon oxynitride based on the overall reaction is 4150 K. According to the above postulation, sand might act as a diluent during the nitriding reaction of silicon and so it can reduce this adiabatic temperature to a value of 3300K instead. However, this temperature is still higher than the decomposition temperatures of either Si_3N_4 (~ 2250 K) or Si_2N_2O (~ 3225 K) as calculated under the experimental nitrogen pressure of 3 MPa. This result explains the presence of residual melted silicon in the product. Figure 6 shows the variation of the adiabatic temperature of the combustion reaction with the Si_2N_2O additions.

The real combustion temperatures, T_{max}, of the reactions as measured by the W/Re thermocouple were much lower than the calculated adiabatic temperatures and did not differ much with the Si_2N_2O additions (Fig. 7). This result was observed earlier on the work of Hirao et al[29] during the nitridation of silicon in presence of Si_3N_4 additions for the synthesis of Si_3N_4. They attributed this to the endothermic decomposition of Si_3N_4 to Si and N at high temperatures. The combustion temperature therefore does not exceed a characteristic temperature, T_{max}, determined by the equilibrium between the exothermic formation reactions and the endothermic dissociation of Si_3N_4.

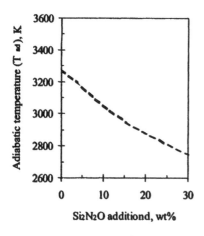

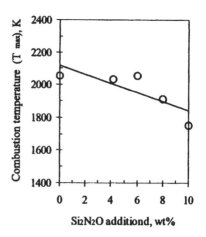

Figure 6. Variation of calculated adiabatic temperature, T_{ad}, with Si_2N_2O additions

Figure 7. Variation of combustion temperature, T_{max}, with Si_2N_2O additions

Thermograms of Combustion Reaction

The thermogram of the nitriding combustion-synthesis reaction clearly explains the kinetics of the reaction. First, if we consider the thermogram of the nitriding combustion-synthesis reaction of reclaimed silicon alone under the same experimental nitrogen pressure, see Fig. 8. The combustion temperature increases rapidly (in ~ 40 s) to T_{max} (~ 2110 K) and is maintained 10 s then decreases suddenly to a temperature of ~ 1530 K followed by a gradual decrease. This profile suggested that silicon was partially nitrided in the combustion front and the nitridation reaction was completed 10 s after the combustion wave had passed. For the thermogram of the nitriding combustion synthesis of Si_2N_2O from the mixture of reclaimed Si, sand, and 10 wt % Si_2N_2O under 3 MPa nitrogen gas (Fig. 8), it had a close trend. The combustion temperature took only 20 s to reach T_{max} and the reaction continued for ~ 90 s after the passage of the combustion wave having characteristic downs and ups for the temperature during this stage. These downs and ups in the combustion temperature during the stage of reaction completion confirmed that there are two reactions with different exothermicities and rates were occurring, and this results confirms the two-steps mechanism postulation. The higher heat release of this reaction during the passage of the combustion front shows that the two steps partially occurred forming Si_2N_2O on the combustion front. T_{max} didn't differ in either case and reached close values below the decomposition of Si_3N_4.

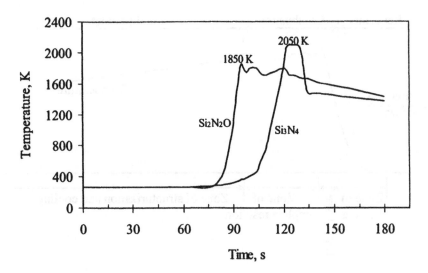

Figure 8. Thermograms of the combustion synthesis reaction of Si_3N_4 and Si_2N_2O under 3 MPa nitrogen pressure

Primary Structure of the Combustion Wave

In conclusion, the nitriding combustion-synthesis of Si_2N_2O from a mixture of reclaimed silicon and desert sand occurred in a two steps synthesis reaction and can be explained by the following simple model, Fig. 9. The synthesis process can be considered in nonequilibrium adiabatic conditions and comprised of several zones; zone of heating, reaction zone, reaction completion zone, and zone of cooling and structure formation in the product. When the initial reactants enter the zone of heating, silicon particles are rapidly heated and begin to react with nitrogen forming silicon nitride phase and producing rapid increase in the temperature which spontaneously activates a chemical combination of the formed nitride phase and sand to form silicon oxynitride phase. The synthesis wave propagates and moves before the completion of the reaction. The decomposition of silicon nitride begins as the temperature approaches the decomposition temperature of silicon nitride so the temperature rises to T_{max} below the decomposition temperature of silicon nitride.

POROUS SILICON OXYNITRIDE

The resultant silicon oxynitride powder was mixed with 5 wt% calcia-alumina additive by the planetary mill for 1h in ethanol using agate milling balls. After

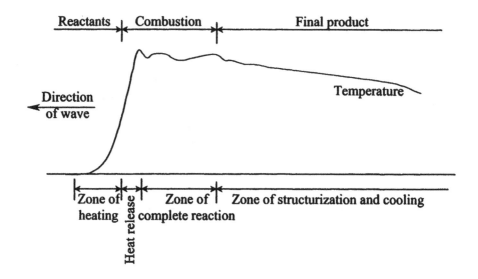

Figure 9. Structure of combustion synthesis wave under non-equilibrium condition

drying, ten grams of the dried powder mixture was pressed into a 30 mm diameter compact by cold isostatic pressing at 200 MPa. Then, the green compacts were heated at a temperature of 1500°C for 1 h in a nitrogen atmosphere. The heating and the cooling rates were 20 °C/min. This resulted in a compact having a bulk density of 1850 kg/m^3 which corresponds to 34.1 % porosity.

To investigate the chemical stability for oxidation, 0.2 g porous specimens (5 × 5 × 10 mm) were heated to 1500°C with a heating rate of 5 °C/min in an oxidizing atmosphere composed of Ar/O$_2$/H$_2$O (70/20/10 kPa). The results showed that the sample retained its shape. Fig. 10 shows the weight gain during the heating test. The specimen showed no weight gain until 1100 °C, then had a little weight gain of 2 mg/cm^2. The surface of the specimen turned whitish.

SINTERING OF SILICON OXYNITRIDE

Green compacts were prepared with the same procedure described above. The sintering experiments were carried out using Dr. Sinter® Model 1050 SPS apparatus (Sumitomo Coal Mining Company, Ltd., Japan). The green compact was placed inside 30 mm diameter graphite die and BN powder was used to surround this compact. The temperature was increased up to 1600 °C with a heating rate of 400 °C/min. The sintering is kept for 2 minutes under a pressure of

30 MPa which was applied from the beginning of the heating. The cooling rate was the same as the heating rate.

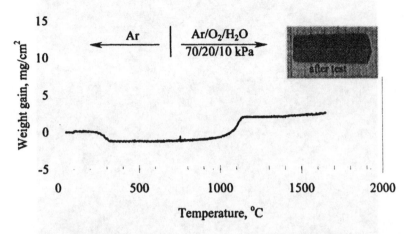

Figure 10. Weight gain of a porous Si_2N_2O compact during heating to 1500 °C in $Ar/O_2/H_2O$ atmosphere (70/20/10 kPa). The photo in the upper right corner shows the compact after the test

The sintered compacts were approximately 28 mm in diameter and 5 mm thick. The properties of the dense Si_2N_2O are summarized in Table I.

Table I. Properties of the sintered Si_2N_2O

Density, %	99.0
Vickers hardness, GPa	18.7
Fracture toughness, MPa m$^{1/2}$	3.3
Flexural strength, MPa	363.0

The sintered compact had a density value of 99 % of the theoretical density and had comparable mechanical properties to the conventional Si_2N_2O materials. The Vickers hardness and fracture toughness of the sintered sample are 18.7 GPa and 3.3 MPa m$^{1/2}$, respectively. In the literature, hardness and fracture toughness values for dense silicon oxynitride were 15-22 GPa and 2.5-6 MPa m$^{1/2}$, respectively[13,18,19]. The three-point flexure strength measured at room temperature is 363 MPa and the reported values in the literature were 300-750 MPa.[13,16,18,19]

Figure 11 shows the micrograph for the fractured surface. The structure has equiaxed grains and the fracture occurred in a transgranular mode.

Figure 11. SEM micrograph of the fractured surface of dense Si_2N_2O

The oxidation resistance results of the sintered samples heated in dry air at 1200°, 1400° and 1500 °C for 10 h are shown in Table II. The sintered material showed excellent resistance at the temperatures up to 1500 °C. Room temperature chemical resistance to the solutions 1M sulfuric acid, 2M sodium hydroxide and 2M sodium chloride are almost infinitive for soaking periods reached 200h.

Table II. Weight gain due to oxidation resistance of dense Si_2N_2O

Test condition	Weight gain, mg/cm^2
1200 °C, 10 h	0.387
1400 °C, 10 h	0.641
1500 °C, 10 h	1.000

SUMMARY

This study is to extend the potential applications of desert sand to synthesize silicon-based ceramics to be used as refractories and corrosion resistance ceramics. Pure silicon oxynitride powder has been successfully prepared from a mixture of desert sand and reclaimed silicon by using the nitriding combustion-synthesis method. Porous Si_2N_2O can wistand at severe environment at high temperature of 1500 °C. Near full dense Si_2N_2O has been prepared at 1600 °C by the spark plasma sintering method and the material had mechanical properties comparable to the reported values.

ACKNOWLEDGEMENT

M. Radwan sincerely appreciates the Missions Department at the Ministry of Higher Education of Egypt for the financial support of his stay in JWRI during this study.

REFERENCES

[1]W. C. Schumb and R. A. Lefevre, "Ammonolysis of Hexachlorodisiloxane", *Journal of The American Chemical Society*, **76**, 5882-84 (1954).

[2]W. D. Forgeng and B. F. Decker, "Nitrides of Silicon", *Transactions of the Metallurgical Society of AIME*, 212 [3] 343-48 (1958).

[3]C. Brosset and I. Idrestedt, "Crystal Structure of Silicon Oxynitride, Si_2N_2O", *Nature*, **201** [4925] 1211 (1964).

[4]K. H. Jack, "Sialons: A Study in Materials Development", pp. 1-30 in *Non-Oxide Technical and Engineering Ceramics*, Edited by S. Hampshire. Elsevier Applied Science Publishers Ltd., London, 1986.

[5]M. B. Trigg and K. H. Jack, "Solubility of Aluminium in Silicon Oxynitride", *Journal of Materials Science Letters*, **6**, 407-8 (1987).

[6]M. E. Washburn, "Silicon Oxynitride Refractories", *Ceramic Bulletin*, **46** [7] 667-71 (1967).

[7]M. E. Washburn, "Production of Silicon Oxynitride", U.S. Pat. No. 3 356 513, Dec. 5, 1967.

[8]M. E. Washburn, "Producing Silicon Oxynitride", U.S. Pat. No. 3 639 101, Feb. 1, 1972.

[9]M. E. Washburn and S. D. Hartline, "Lightweight Silicon Oxynitride", U.S. Pat. No. 4 043 823, Aug. 23, 1977.

[10]M. E. Washburn, "Porous Silicon Oxynitride Refractory Shapes", U.S. Pat. No. 4 069 058, Jan. 17, 1978.

[11]M. E. Washburn, "High Density Silicon Oxynitride", U.S. Pat. No. 4 331 771, May 25, 1982.

[12]P. Boch and J. C. Glandus, "Elastic Properties of Silicon Oxynitride", *Journal of Materials Science*, **14**, 379-85 (1979).

[13]M. H. Lewis, C. J. Reed and N. D. Butler, "Pressureless-Sintered Ceramics Based on the Compound Si_2N_2O", *Materials Science and Engineering*, **71**, 87-94 (1985).

[14]M. Komatsu and I. Ikeda, "Sintered Ceramic Articles and Method for Production Thereof", U.S. Pat. No. 4 761 339, Aug. 2, 1988.

[15]H. Yokoi, S. Iio and S. Watanabe, "Sintered Body of Silicon Oxynitride", JP Pat. No. 1 264 972 A2, Oct. 23, 1989.

[16]M. Mitomo, S. Ono, T. Asami and Suk-Joong L. Kang, "Effect of Atmosphere on the Reaction Sintering of Si_2N_2O", *Ceramics International*, **15**, 345-50 (1989).

[17]S. Kanzaki, M. Ohashi, H. Tabata, O. Abe, T. Himamori and K. Moori, "Sintered High Density Silicon Oxynitride and Method for Making the Same", U.S. Pat. No. 4 975 394, Dec. 4, 1990.

[18]M. Ohashi, S. Kanzaki and H. Tabata, "Processing, Mechanical Properties, and Oxidation Behavior of Silicon Oxynitride Ceramics", *Journal of The American Ceramic Society*, **74** [1] 109-14 (1991).

[19]R. Larker, "Reaction Sintering and Properties of Silicon Oxynitride Densified by Hot Isostatic Pressing", *Journal of The American Ceramic Society*, **75** [1] 62-66 (1992).

[20]P. W. Lednor, "Synthesis, Stability, and Catalytic Properties of High Surface Area Silicon Oxynitride and Silicon Carbide", *Catalysis Today*, **15**, 243-61 (1992).

[21]W. R. Ryall and A. Muan, "Silicon Oxynitride Stability", *Science*, **165**, 1363-1364 (1969).

[22]T. C. Ehlert, T. P. Dean, M. Billy and J.-C Labbe, "Thermal Decomposition of the Oxynitride of Silicon", *Journal of The American Ceramic Society*, **63** [3-4] 235-36 (1980).

[23]M. Bruce Fegley, JR., "The Thermodynamic Properties of Silicon Oxynnitride", *Journal of The American Ceramic Society*, **64**, C-124-26 (1981).

[24]M. Ekelund, B. Forslund, G. Eriksson and T. Johansson, "Si-C-O-N High-Pressure Equilibria and ΔG^o_f for Si_2ON_2", *Journal of The American Ceramic Society*, **71** [11] 956-60 (1988).

[25]M. B. Trigg, "Method of Forming a Ceramic Product", U.S. Pat. No. 4 987 104, Jan. 22, 1991.

[26]P. W. Lednor, "Process for the Preparation of Silicon Oxynitride-Containing Products", U.S. Pat. No. 4 977 125, Dec. 11, 1990.

[27]A. G. Merzhanov, "Combustion Processes that Synthesize Materials", *Journal of Materials Processing and Technology*, **56**, 222-41 (1996).

[28] Y. Miyamoto, "State of the Art in the R&D of SHS Materials in the World", *International Journal of the Self-Propagating High-Temperature Synthesis*, **8** [3] 375-84 (1999).

[29]K. Hirao, Y. Miyamoto and M. Koizumi, "Combustion Reaction Characteristics in the Nitridation of Silicon", *Advanced Ceramic Materials*, **2** [4] 780-83 (1987).

FABRICATION AND EVALUATION OF POROUS Ca-α SiAlON CERAMICS

Junichi Tatami, Takayuki Ohta,
Cheng Zhang, Mikinori Hotta,
Katsutoshi Komeya
and Takeshi Meguro
Yokohama National University,
79-7 Tokiwadai Hodogayaku
Yokohama 240-8501, Japan

Mamoru Omori and Toshio Hirai
Institute for Materials Research
Tohoku University, Sendai
2-1-1 Katahira, Aobaku, Sendai
980-8577, Japan

ABSTRACT

We previously reported that 200 to500nm in diameter Ca-α SiAlON hollow ball consisting of very fine 10 to 30nm in diameter particles could be synthesized by the carbothermal reduction nitridation method. In this study, we investigated fabrication of porous Ca-α ceramics by using the hollow balls and maintaining their unique shape. Experimental results revealed that spark plasma sintering (SPS) effectively densify nano-sized Ca-α SiAlON powder. We also attempted to sinter the hollow balls by SPS at 1700°C, 1 to 20 min, in N_2 under 5MPa uniaxial pressure. The open and closed porosity of the sample fired at 1700 °C for 1 min is 44 and 1%, respectively. XRD analysis indicated that phases present are almost the same between raw material and porous ceramics. SEM observation showed that the hollow shape of the raw material was maintained after SPS in the sample of 1min soaking, though the pore was somewhat squashed in the sample of 20min soaking. The developed porous Ca-α SiAlON ceramics has high strength, high hardness and high corrosion resistance. Consequently, we confirmed that the SPS method was advantageous for fabricating porous Ca-α SiAlON ceramics.

INTRODUCTION

α- and β- SiAlON are solid solutions of α- and β- Si_3N_4, in which Si and N are replaced with Al and O, respectively. Furthermore, cations, such as Li^+, Mg^{2+}, Ca^{2+}, Y^{3+} and some rare earth ions, are included in the α-SiAlON structure.[1] α-SiAlON is characterized by high heat resistance, good corrosion resistance and high hardness. Recently, α-SiAlON was strengthened and toughened by elongating the grain structure [2, 3].

In our previous study on α-SiAlON, we synthesized Ca-α SiAlON powders by carbothermal reduction-nitridation of a SiO_2-Al_2O_3-$CaCO_3$ powder mixture. As a result, we obtained Ca-α SiAlON hollow balls composed of nano particles[4]. After grinding them, we developed dense Ca-α SiAlON ceramics with higher corrosion resistance than conventional Si_3N_4 ceramics.[5] If we sintere the hollow balls by maintaining their unique morphology, we expect that novel porous ceramics with high corrosion resistance can be developed. The purpose of this study is to develop porous SiAlON ceramics by using Ca-α SiAlON hollow balls and evaluate their microstrucutre, mechanical properties and corrosion resistance.

EXPERIMENTAL PROCEDURE
The starting powders were SiO_2 (>99.99%, QS-102, Tokuyama Co., Ltd.), Al_2O_3 (>99.99%, AKP50, Sumitomo Chemical Co., Ltd.), $CaCO_3$ (>99.0%, Junsei Chemical Co. Ltd.) and carbon (>99.97%, 650B, Mitsubishi Chemical Co. Ltd.). $CaCO_3$ in the mixture decomposes above 900°C, producing CaO and CO_2. We weighed these powders to yield nominal SiAlON compositions of $Ca_xSi_{12-3x}Al_{3x}O_xN_{16-x}$ (x=0.8-1.4), and fixed the carbon content to 1.2 times the required stoichiometric value. We then milled the mixtures with ethanol in an agate mortar for 120min, ground the dried powders, passed them through a sieve with 300μm holes and put the mixed powders into a graphite boat. Firing was performed in a horizontal electrical furnace at 1450°C for 120min. We employed a constant nitrogen gas flow of 0.5l/min through the furnace during the whole heating cycle and removed the residual carbon by burning the synthesized powder at 700°C for 120min in air. In this way, we confirmed that the synthesized powders exhibited the shape of hollow balls composed of nano sized particles by SEM and XRD. Using the powders, we performed spark plasma sintering (SPS) at 1700°C for 1 to 20 min under uniaxial pressure of 5MPa in 0.1MPa N_2. Hot pressing (HP), Gas pressure sintering (GPS) and induction heating sintering (IHS) were also carried out using the same powder for reference. We evaluated density by Archimedes method, pore size distribution by mercury porosimeter and phases present by X-ray diffraction (XRD) analysis. We observed the microstructure by scanning electron microscopy (SEM), measured the bending strength a 3-point bending test, and estimated hardness the Vickers indentation method. Corrosion resistance was calculated from the weight change after a specimen was put in 80°C 5wt% H_2SO_4 for 80 hours.

RESULT AND DISCUSSION
Figure 1 shows the XRD profile and the SEM photograph of the synthesized raw powder. The powder consisted of hollow balls with a diameter of about 500nm composed of nano-sized particles, and the phase present was mainly

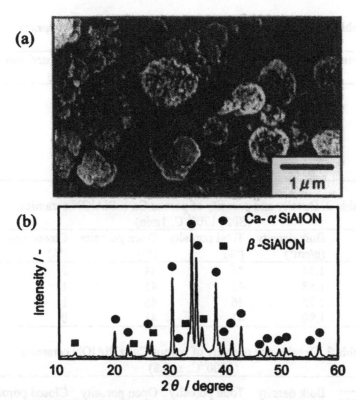

(a)

1 μm

(b)

● Ca-α SiAlON

■ β-SiAlON

Intensity / -

2θ / degree

Fig. 1 Ca-α sialon hollow balls synthesized by carbothermal reduction
-nitridation of SiO_2-Al_2O_3-$CaCO_3$
(a) SEM photograph and (b) XRD profile

Ca–α SiAlON. We thus confirmed that the size, shape, and phase of the present product are identical to the hollow balls produced in the previous paper.

Table 1 shows the density and porosity of the products fired at a temperature of 1700°C for various soaking times in spark plasma sintering. We used 3.228g cm^{-3} $^{6)}$ as true density to calculate the relative density and the porosity. Density increased and porosity decreased with an increase in the soaking time in spite of a small dispersion during soaking. Furthermore, almost all pores in the products were found to be open. Table 2 also presents the density and porosity of the porous ceramics sintered from the powder with different x values in $Ca_xSi_{12-3x}Al_{3x}O_xN_{16-x}$. These samples had almost the same density and porosity, and almost all pores were also open. As shown in Table 3, the samples sintered by HP, GPS and IHS had lower density than those sintered by SPS. Considering the green density (1.40 to 1.42(g cm^{-3})), the green body shrank very little HIS, athough

Table 1. Density and porosity of porous Ca-α SiAlON ceramics (SPS, 1700°C, x=0.8)

Soaking time (min)	Bulk density (g/cm^3)	Total porosity (%)	Open porosity (%)	Closed porosity (%)
1	1.74	46	44	2
1	1.88	42	41	1
6	2.32	28	27	1
20	2.35	27	25	2
20	2.51	22	21	1
20	2.63	19	17	2

Table 2. Density and porosity of porous Ca-α SiAlON ceramics (SPS, 1700°C, 1min)

x value	Bulk density (g/cm^3)	Total porosity (%)	Open porosity (%)	Closed porosity (%)
0.8	1.74	46	44	2
0.8	1.88	42	41	1
1.0	1.75	46	45	1
1.4	1.90	41	41	0

Table 3. Density and porosity of porous Ca-α SiAlON ceramics (1700°C, x=0.8)

Method / Soaking time (min)	Bulk density (g/cm^3)	Total porosity (%)	Open porosity (%)	Closed porosity (%)
HP / 1	1.46	55	54	1
GPS / 20	1.47	55	54	1
GPS / 60	1.46	54	54	0
IHS / 1	1.01	69	—	—
HIS / 20	1.08	67	—	—

the samples through HP and GPS were slightly densified. Therefore, it is confirmed that the green body composed of Ca–α SiAlON hollow balls was temperate in sintering by SPS than in sintering by HP, GPS and IHS.

Figure 2 depicts XRD profiles of the developed porous ceramics. In all

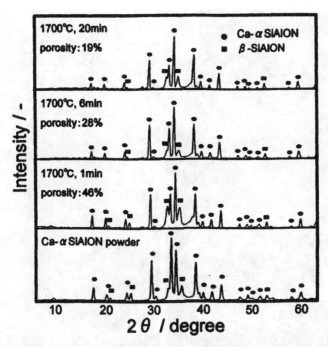

Fig. 2 XRD profiles of the product

samples, the main phase was Ca–α SiAlON and the secondary phase was small amounts of β-SiAlON and aluminum nitiride, which were almost the same as Ca-α SiAlON powder (Fig. 1(a))

Figure 3 presents SEM photographs of the fracture surface of the porous Ca–α SiAlON ceramics fabricated by SPS. In the sample with high porosity, we observed that hollow balls with the same morphology as the starting powder (Fig. 2(a)) remain after sintering. Furthermore, we were able to show that the balls gradually disappear as soaking time increases. Figure 4 illustrated the microstructure of the samples sintered by HP and GPS. Although some hollow balls remained after sintering, we observed remarkable grain growth, which corresponded to the sharpening of the XRD peak of Ca–α SiAlON. As a result, we found that HP and GPS which have relatively low heating rate, are unsuitable methods of fabricating porous ceramics from Ca–α SiAlON hollow balls because of grain growth during sintering.

Figure 5 shows the pore-size distribution of the porous Ca–α SiAlON ceramics produced by SPS. The porous material with a porosity of 46% had a large number of pores with diameters of 200 to 500nm and a small number of

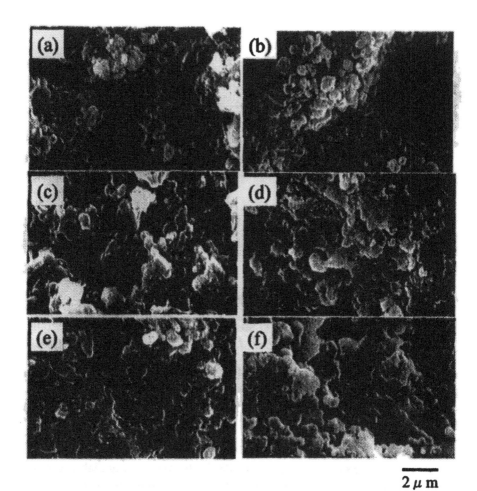

2 μ m

Fig. 3 SEM photographs of the fracture surfaces of porous materials fabricated by SPS fired at 1700°C for various soaking time.
(a)1min (46% porosity), (b)1min (42% porosity), (c) 6min (28% porosity)、
(d)6min (27% porosity), (e)20min (22% porosity), (c)20min (19% porosity)、

pores with diameters of about 10nm. There were also of 200-500nm diameter pores in the sample with a porosity of 27 to 28%, but these were fewer than that with a porosity of 46%. The 200 to 500nm diameter pores were not observed in the sample with a porosity of 18%. Considering the size of raw Ca–α SiAlON hollow balls, the balls and the nano-sized primary particles seems to generate necking with a small amount of shrinkage during SPS in the high porosity ceramics to produce the pores. It is thus probable that sintering progressed and

Silicon-Based Structural Ceramics

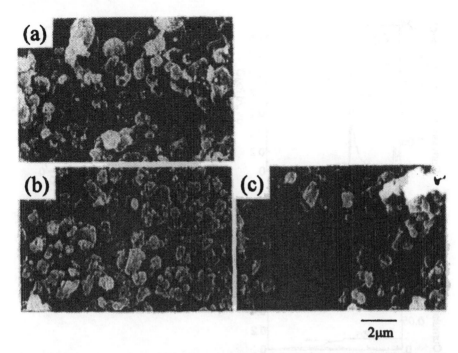

2μm

Fig. 4 SEM photographs of the fracture surfaces of porous materials fabricated by HP and GPS under various conditions. (a)HP 1700°C, 1min (55% porosity), (b)GPS 1700°C, 1min (55% porosity), (c)GPS 1700°C, 1min (54% porosity),

pores shrunk as soaking time increased.

Figure 6 describes the relationship between the porosity and the strength of porous Ca–α SiAlON ceramics. The strength decreased with an increase in porosity. Generally speaking, pores in the material reduce the effective load-bearing area, thus degrading Young's modulus. Furthermore, a pore usually behaves as a fracture origin. As a result, an increase of porosity degrades the strength. Duckworth derived the following equation for the relationship between strength and porosity from the strength data for several ceramics:[7]

$$\sigma = \sigma_0 \cdot \exp(-bP) \tag{1}$$

where σ and σ_0 indicate the strength of the sample with porosity of P and 0, and b is a constant. This equation is confirmed to be experimentally valid for many kinds of ceramics.[8-14] The solid line in the Fig. 5 is the regression curve calculated from experimental data by using equation (1). This curve demonstrates

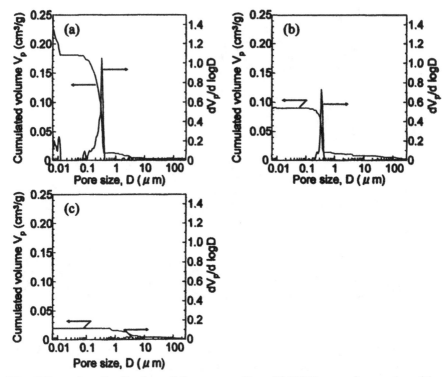

Fig. 5 Pore-size distribution of the porous Ca-α SiAlON ceramics produced by SPS. (a)1700°C, 1min, (b)1700°C, 6min, (c)1700°C, 20min

that our result can also be expressed very well by using equation (1). Figure 6 also indicates the strength of porous alumina ceramics reported by Coble.[8] The strength of the newly developed Ca–α SiAlON ceramics in this study can be seen to be much higher than that of the former and typical porous ceramics. Furthermore, the strength of the porous Ca–α SiAlON ceramics with a porosity of 20% is almost the same as both that of dense mullite and cordierite, which are used for filters, and that of reaction-bonded silicon nitride (RBSN, porosity 22%)[15].

Figure 7 shows the relationship between the porosity and the Vickers hardness of porous Ca–α SiAlON ceramics. Hardness as well as strength decreased with an increase in porosity though it is difficult to analyze the hardness of porous material because it includes the effect of quasi-plastic deformation due to microscopic fracture arising from compressive and shear stress under indentation. The hardness of the porous Ca–α SiAlON ceramics was higher than that of dense

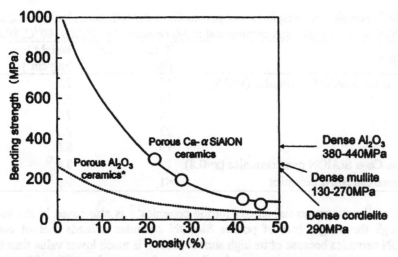

Fig. 6 The relationship between the porosity and the strength of porous Ca-α SiAlON ceramics.
*R. L. Coble et. al., J. Am. Ceram. Soc., 39, 11, 377-85 (1956).

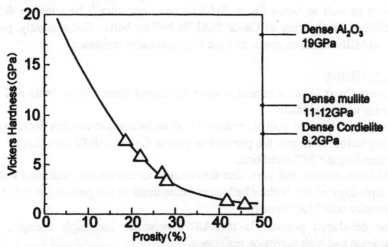

Fig. 7 The relationship between the porosity and the hardness of porous Ca-α SiAlON ceramics.

mullite and cordierite and almost the same as that of RBSN[15] and porous silicon nitiride ceramics fabricated by using starch.[16]

Table 4 lists the weight loss and change in density after corrosion testing in 80°C 5wt% H_2SO_4 for 80h. The corrosion test data of dense Ca-α SiAlON nano

Table 2. Porosity and weight loss of porous Ca-α SiAlON ceramics, dense Ca-a SiAlON nano ceramics and commercial Si_3N_4 ceramics ($5\%H_2SO_4$, 80^oC, 80h)

Sample	Total porosity (%)	Weight loss (mg/cm^2)
Porous Ca-α SiAlON ceramics (x=0.8)	46	7.7
	42	5.6
	28	7.7
	22	7.1
	19	8.0
Dense Ca-α SiAlON nano ceramics (x=0.8)	1	3.6 [5]
Commercial Si_3N_4 ceramics	<1	9.5 [17]

ceramics [5] and commercial silicon nitride ceramics [17] is also listed in the table. Although the weight loss of porous SiAlON ceramics exceeds that of dense SiAlON ceramics because of its high surface area, it is much lower value than that of commercial silicon nitiride ceramics. In general, glassy phase formed by phase reaction of SiO_2 and sintering aids (Y_2O_3, Al_2O_3 and so on) also exists in the grain boundary of silicon nitride ceramics. It has been reported that the glassy phase dissolves easily in acid.[17)18)] However, there may be glass phase in the grain boundary as well as dense Ca–α SiAlON nano ceramics [19] because we did not use additive for sintering of Ca–α SiAlON hollow balls . Consequently, porous Ca–α SiAlON ceramics seems to have high corrosion resistance.

CONCLUSIONS

Porous Ca-α SiAlON ceramics were fabricated from hollow balls and their properties were evaluated.

1. Porosity of sintered bodies decreased with an increase in soaking time, so that it is possible to control the porosity of porous Ca-α SiAlON ceramics by controlling the SPS condition.

2. SEM observation and pore size distribution measurement confirmed that the morphology of the hollow balls was maintained in the porous Ca-α SiAlON ceramics with high porosity.

3. The developed porous Ca-α SiAlON ceramics had high strength, high hardness and high corrosion resistance.

REFERENCES

[1] S. Hamphire, H. K. Park, D. P. Thompson and K. H. Jack, "α'-SiAlON," Nature (London), 274, 880-82 (1978).

[2] I-Wei Chen and A. Rosenflanz, "A Tough SiAlON Ceramic Based on Alpha Si3N4 with a Whisker-like Microstructure," Nature, 389, 701-04 (1997).

[3] C. A. Wood, H. Zhao and Y. B. Cheng, "Microstructural Development of Ca

a-SiAlON Ceramics with Elongated Grains," J. Am. Ceram. Soc., 82 [2], 421-28 (1999).

[4]K. Komeya, C. Zhang, M. Hotta, J. Tatami, T. Meguro and Y. -B. Cheng, "Hollow Beads Composed of Nano-Size Ca-α Sialon Grains," J. Am. Ceram. Soc., 183[4], 995-97(2000).

[5]J. Tatami, M. Iguchi, M. Hotta, C. Zhang, K. Komeya, T. Meguro, M. Omori, T. Hirai, M. E. Brito and Y.-B. Cheng, "Fabrication and Evaluation of Ca-α SiAlON Nano Ceramics", Key Engineering Materials, 237, 105-110 (2002).

[6]JCPDS card No. 33-0261

[7]W.H. Duckworth, "Discussion of Ryshke-witch Paper by Winston Duckworth," J. Am. Ceram. Soc. 36, 68 (1953).

[8]R. L. Coble and W. D. Kingery, "Effect of Porosity on Physical Properties of Sintered Alumina," J. Am. Ceram. Soc., 39 [11] 377-85 (1956).

[9]W. D. Kingery , "The Physics and Chemistry of Ceramics" (ed. by C. Klingsberg), Gordon and Breach, New York, 286 (1963).

[10]F. P. Knudsen, "Dependence of mechanical strength of brittle polycrystalline specimens on porosity and grain size", J. Am. Ceram. Soc., 42(8), 376-387 (1959).

[11]R. L. Studt, R. H. Fulrath, J. Am. Ceram. Soc., 45, 182-88 (1962).

[12]L. J. Trostel, Jr., J. Am. Ceram. Soc., 45, 563-64 (1962).

[13]R. M. Spriggs, T. Vasilons, J. Am. Ceram. Soc., 46, 224-28 (1963).

[14]R. E. Fryxell, B. A. Chandler, J. Am. Ceram. Soc., 47, 283-91 (1964).

[15]Handbook of Ceramics [2nd Edition], Edited by Ceramic Society of Japan, Gihodo, Tokyo, 2002.

[16]A. Diaz and S. Hampshire "Fracture of Porous Ceramics", in Proceedings of the 6th Symposium on Synergy Ceramics, Tokyo, Japan, 22-23, (2002).

[17]K. Komeya, T. Meguro, S. Atago, C.-H. Lin, and M. Komatsu, "Corrosion Resistance of Silicon Nitride Ceramics," Key Engineering Material, 161-163, 235-239 (1999).

[18]T. Sato, Y. Tokunaga, T. Endo, M. Shimada, K. Komeya, K. Nishida, K. Komatsu and T. Kameda, "Corrosion of Silicon Nitride Ceramics in Aqueous Hydrogen Chloride Solutions", J. Amer. Ceram. Soc., 71, 1074-1079 (1988).

Microstructures: Development and Characterization

HIGH RESOLUTION IMAGING AND MICROANALYSIS OF SILICON-BASED CERAMICS

Lena K. L. Falk
Department of Experimental Physics
Chalmers University of Technology
SE-412 96 Göteborg
Sweden

ABSTRACT

This paper is focussed on the characterization of microstructural development during liquid phase sintering and post-densification heat treatment of ceramic materials based on the Si_3N_4 or SiC structures. The microstructures, in particular the intergranular regions, have been characterized by high resolution electron optical techniques for imaging and microanalysis, and are related to the over-all composition and different fabrication parameters. It is demonstrated that combined high resolution analytical and spatial information from chemically and structurally distinct fine scale features, such as grain boundary films of residual glass, is obtained by electron spectroscopic imaging and subsequent computation of elemental distribution images.

INTRODUCTION

The SiC and α- and β-Si_3N_4 structures contain small and comparatively closely packed atoms. The interatomic bonding in these structures is strongly covalent, which, together with the atomic arrangement, gives a large number of strong bonds per unit volume. This results in a combination of good inherent mechanical and chemical properties such as a high strength, a high value of Young's modulus, good oxidation and corrosion resistance and a low thermal expansion. Ceramic materials based on the SiC or Si_3N_4 structures are, consequently, potential high strength materials for structural applications at both high and ambient temperatures [1-7].

The strong interatomic bonding results, however, also in extremely low self-diffusivities at temperatures below that where appreciable decomposition of the ceramic compound starts. The fabrication of dense SiC and Si_3N_4-based materials requires, therefore, a sintering additive that promotes densification, and at the same time inhibits the decomposition of the ceramic compound [1-4, 7-11].

The sintering additives will, in general, introduce an intergranular microstructure that may limit the performance in an application [1-7, 12-18]. Refractory secondary phases and a strong intergranular bonding are required for high strength and creep resistance at elevated temperatures. The activation of different toughening mechanisms such as crack deflection and bridging, and micro-cracking rely, on the other hand, on "sufficiently weak" interfaces. It becomes, hence, important to control the intergranular microstructure, in addition to grain morphology and composition.

LIQUID PHASE SINTERING

Liquid phase sintering of ceramic materials based on the SiC or Si_3N_4 structures requires the addition of metal oxides or nitrides [1-4, 7, 9, 19-20]. Above relevant eutectic temperatures, the sintering additives react with the inherent surface silica present on the starting powder particles, and some of the SiC or Si_3N_4, whereby a liquid phase sintering medium is formed. The chemistry and the volume fraction of the liquid is determined by the choice of additives as well as by densification parameters such as the atmosphere and the temperature / time program during sintering. Densification is achieved through particle rearrangement in the liquid phase accompanied by a solution / reprecipitation process.

The final microstructure consists of SiC / Si_3N_4 or sialon grains, and, in general, also intergranular phases, see Figure 1. Secondary crystalline phases may partition from the liquid phase sintering medium, but a certain volume fraction of the liquid is generally retained as a residual glass in the microstructure of the sintered ceramic.

Impurities that originate from the starting powders, or are introduced during the preparation of green bodies, will concentrate to the liquid phase sintering medium, and eventually be part of the intergranular microstructure. This may have a detrimental effect on the high temperature properties of the ceramic because of the less refractory nature of impurity containing glass and secondary phases.

IMAGING AND ANALYSIS
Grain Shape and Size

The properties of the liquid will determine the densification rate and the extent of grain growth, and thereby the grain shape and size distribution of the sintered material. The overall microstructure may be described by two-dimensional grain section parameter distributions determined by quantitative microscopy. The application of stereological methods are, however, needed for the assessment of a true three-dimensional grain size distribution from experimentally determined two-dimensional parameter distributions, and these methods require, in general, an hypothesis on grain shape [21].
Residual Glass

Larger volumes of residual glass are readily identified in the transmission electron microscope (TEM) because of their lack of diffraction contrast when the

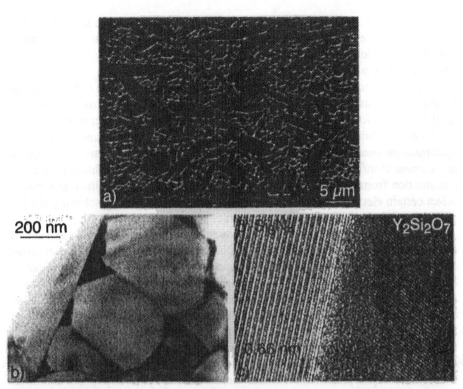

Figure 1: Silicon nitride based microstructures fabricated with additions of Y_2O_3 and Al_2O_3. (a) Plasma etched section displaying prismatic β' grains separated by an intergranular microstructure that appears with bright contrast (SEM). (b) Partly crystallized intergranular microstructure where the secondary crystalline phase is separated from the β' grains by residual glass as shown in (c).

thin foil specimen is tilted under the electron beam. This technique is, however, not applicable when the glass is present as thin grain boundary films.

Grain boundary structures and intergranular films can be directly, as well as indirectly, identified and imaged by well established techniques in the TEM [22-26]. Dark field images formed from part of the diffuse scattering from the glass are direct images of the distribution of the intergranular glass. High resolution lattice imaging makes it possible to image grain boundary detail with the resolution of an interplanar spacing. Defocus Fresnel imaging is an indirect technique for grain boundary film determination. An underfocus image of a grain boundary containing a film of residual glass contains a pair of dark lines (fringes) delineating the boundary. This contrast is reversed when going through Gaussian focus so that overfocus images contain a pair of bright lines delineating the boundary. The spacing between the fringes depends upon the defocus and the film thickness. This makes it possible to estimate grain boundary film thickness from plots of fringe spacing versus defocus.

Chemical Composition

The correlation of chemical information to fine scale structure can be obtained by energy dispersive X-ray (EDX) analysis or electron energy loss spectroscopy (EELS) when a focussed probe is stepped across the thin foil specimen in the TEM.

High resolution chemical information can also be obtained by electron spectroscopic imaging [22, 27-29]. These images combine two-dimensional spatial and elemental information, and are produced by an energy filter that separates the contribution from electrons with different energies. It becomes, then, possible to select certain electron energies for imaging by placing an energy selecting slit in a plane containing a focussed EEL spectrum.

The electron energy loss spectrum above the plasmon peak consists of element characteristic edges superimposed on a background that is rapidly decreasing towards higher energy losses. Two types of elemental distribution images can be obtained from electron energy filtered images: elemental maps and jump ratio maps. An elemental map is computed by subtracting a background image, calculated from two pre-edge images, from a post-edge image containing the element specific signal. Jump ratio maps are computed by dividing a post-edge image by a pre-edge image. Figure 2 shows the computation of a carbon map from electron spectroscopic images recorded around the carbon K edge in the electron energy loss spectrum from a SiC ceramic liquid phase sintered with additions of Y_2O_3 and Al_2O_3. An aluminum jump ratio map of the same area was computed from images recorded around the aluminum $L_{2,3}$ edge as shown in Figure 2.

GRAIN SHAPE AND SIZE OF SILICON NITRIDE BASED CERAMICS

Scanning and transmission electron microscopy has demonstrated that the grain shape in many Si_3N_4-based microstructures may, to a good approximation, be described by an hexagonal prism with a certain aspect ratio (length-to-width ratio) [30]. This shape reflects the hexagonal Si_3N_4 lattice, with the c-axis along the length of the prism, and the prism planes forming on the {10-10} crystal planes.

Different methods for estimating β-Si_3N_4 grain shape, and size, from a two-dimensional section through the microstructure (Figure 1) have been presented in the literature. In some investigations, it has been assumed that the Si3N4 grains in a particular microstructure all have the same aspect ratio, and that this aspect ratio may be calculated from the highest apparent aspect ratios on a section through the material [31]. In other studies, a two-dimensional aspect ratio distribution was determined from measurements of the minimum and maximum Feret diameters on individual grain sections [32].

The assessment of a true three-dimensional grain shape requires, however, access to stereological characteristics of bodies with the assumed shape [21]. The

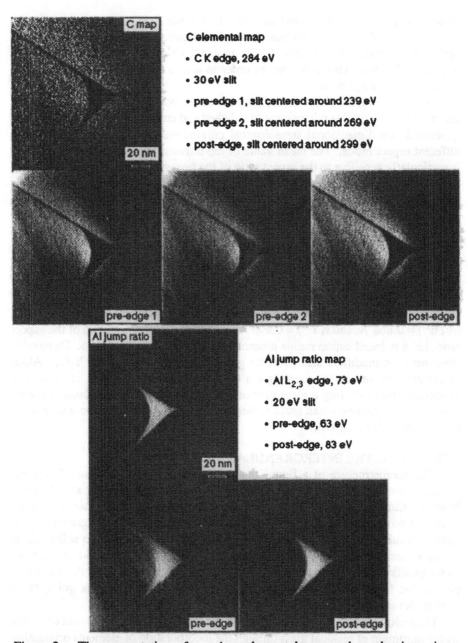

C elemental map

- C K edge, 284 eV
- 30 eV slit
- pre-edge 1, slit centered around 239 eV
- pre-edge 2, slit centered around 269 eV
- post-edge, slit centered around 299 eV

Al jump ratio map

- Al L$_{2,3}$ edge, 73 eV
- 20 eV slit
- pre-edge, 63 eV
- post-edge, 83 eV

Figure 2:   The computation of a carbon elemental map and an aluminum jump ratio map of a triple grain junction in a hot isostatically pressed liquid phase sintered silicon carbide ceramic.

hexagonal prism has been used as a model for the β-Si$_3$N$_4$ grain shape in the determination of three-dimensional grain size distributions of experimental Si$_3$N$_4$ materials formed with a constant total molar fraction of different metal oxide additives [33, 34]. This was done in order to assess the effect of different oxide additives on grain growth.

The average grain shape in the microstructure was determined by comparing results from quantitative microscopy of polished and etched sections with computer-generated two-dimensional stereological parameters of hexagonal prisms with different aspect ratios. The mean value of the dimensionless shape parameter (Q) is particularly sensitive to the aspect ratio of the hexagonal prism, but independent of prism volume, and is therefore a suitable parameter for the determination of an average aspect ratio [21]. Section parameter distributions for the average grain shape determined from the mean value of Q were then obtained from computer simulations and used in a three-dimensional reconstruction of the microstructure as described in reference [21].

The reconstructions showed that oxynitride glass network modifiers promote the development of high aspect ratio β-Si$_3$N$_4$ grains; an increased Y$_2$O$_3$ / Al$_2$O$_3$ molar ratio results in an increased aspect ratio. The work also showed that the radius of the modifying cation affects the aspect ratio. The replacement of Y$_2$O$_3$ by Yb$_2$O$_3$ (r(Yb^{3+}) = 0.868 Å, and r(Y^{3+}) = 0.900 Å) results in a further increase of the aspect ratio, i.e. a reduced cation radius promotes an increase in aspect ratio. Oxynitride glass network modifiers also promote grain growth; increasing the Y$_2$O$_3$ / Al$_2$O$_3$ molar ratio increases the mean grain size of the microstructure. It was also established that selecting a glass network modifying cation with a smaller radius results in an increased mean grain volume and gives a microstructure with a more narrow grain size distribution.

CONTROL OF THE INTERGRANULAR MICROSTRUCTURE

The microstructure of a liquid phase sintered Si$_3$N$_4$- or SiC-based ceramic material contains, in general, an intergranular glassy phase present as thin grain boundary films merging into pockets at multi-grain junctions. This glass is a residue of the liquid phase sintering medium [10, 20, 35]. A reduced intergranular glass volume results when liquid phase constituents are incorporated into solid phases during sintering. This can be achieved through the formation of solid solutions or by the partitioning of secondary crystalline phases. It is important that the secondary phases that form are stable and oxidation resistant if the ceramic is going to be used at elevated temperatures.

The sialon systems have a potential for the fabrication of Si$_3$N$_4$-based ceramics with a minimum of residual glass through the incorporation of elements originating from the sintering additives into the grains of Si$_3$N$_4$ and other, secondary, crystalline phases. The α- and β-sialons are solid solutions based on the hexagonal α- and β-

Silicon-Based Structural Ceramics

Si_3N_4 structures [4, 36-37]. α-sialon (α'-Si_3N_4) has a composition given by $R_x Si_{12-(m+n)}Al_{m+n}O_n N_{16-n}$, where m(Si-N) are substituted by m(Al-N), and n(Si-N) by n(Al-O) in the α-Si_3N_4 structure. The valency discrepancy introduced by this substitution is compensated for by the interstitial cation R^{p+}, and x = m/p. Cations such as Y^{3+}, Yb^{3+}, Dy^{3+}, Sm^{3+} and Nd^{3+} are well known to stabilize the α' structure. The corresponding oxides, as well as Al_2O_3, are efficient liquid phase sintering additives to silicon nitride. β-sialon (β'-Si_3N_4) forms when Si in the β-Si_3N_4 structure is substituted by aluminum, and some nitrogen at the same time is replaced by oxygen for the retention of charge neutrality. The general formula of β-sialon is $Si_{6-z}Al_z O_z N_{8-z}$, where z ≤ 4.2. In addition, the α- and β-sialon structures allow the development of self-reinforced, duplex, microstructures consisting of fibrous β' grains in an α' matrix [39, 40].

Secondary crystalline phases may partition from the liquid during densification, or form during a crystallization heat treatment at a lower temperature [1, 4, 7, 10, 20, 35, 41, 42]. Many of the secondary crystalline phases that form in liquid phase sintered silicon-based ceramics are not stable at the high sintering temperature, and a post-densification heat treatment is therefore required in order to reduce the volume fraction of residual glass. A complete crystallization is, however, generally not obtained even if the chemistry of the starting powder mixture has been carefully designed, see Figures 1 b and c. The liquid and glass phases may support hydrostatical stresses, and these may present an obstacle to the crystallization of smaller liquid or glass volumes [43, 44].

POCKETS OF RESIDUAL GLASS
Fine probe EDX analysis in the TEM has shown that the composition of residual glass pockets is generally not homogeneous [23, 45]. Grain growth during densification results in diffusion profiles and a shift in the average glass composition. This shift in composition results in a glass that is richer in additive cations, and this may promote the formation of secondary crystalline phases during a prolonged holding time at the densification temperature [34, 45]. Elemental concentration profiles across β'- or α'-Si_3N_4 facets into pockets of residual glass have shown that the cations from the sintering additives are anti-correlated with the silicon in the residual oxynitride glass. This has been observed for both glass network formers (Al^{3+}) and network modifiers (Y^{3+}, Sm^{3+}), see Figures 3 and 4. As expected, these profiles show that there is an anti-correlation of aluminum with silicon also in the sialon grains.

The silicon nitride microstructure shown in Figure 3 developed during gas pressure sintering of a powder compact containing a total of 9.5 mol% Y_2O_3 and Al_2O_3 with a Y_2O_3/Al_2O_3 molar ratio corresponding to that of the Y, Al-garnet (YAG), $5Al_2O_3 \cdot 3Y_2O_3$ [34, 45]. A certain amount of aluminum was incorporated into the β-Si_3N_4 lattice, resulting in the formation of a dilute β'-Si_3N_4, but a major

One other example indicating that sialon grain growth is rate controlled by diffusion through the liquid phase sintering medium is shown by the elemental profiles in a samarium α-sialon microstructure in Figure 4. The growing α'-Si₃N₄ grain incorporates some samarium, and has an average aluminum cation fraction that is only marginally higher than that of the residual glass pocket. This results in a significantly increased concentration of samarium, accompanied by a slight reduction in the aluminum concentration, in the glass just outside the moving interface. It may also be noted that the interstitial cation in the α'-Si_3N_4 structure (Sm^{3+}) seems to be anti-correlated with silicon.

FORMATION OF SECONDARY CRYSTALLINE PHASES

The degree of crystallization during the formation of secondary crystalline phases depends upon the composition of the sintering additives as well as the temperature / time program during the densification or a post-densification heat treatment. Work has also shown that parameters such as the atmosphere during sintering and heat treatment, and an applied pressure during densification, may have an effect on the intergranular microstructure [20, 46, 47].

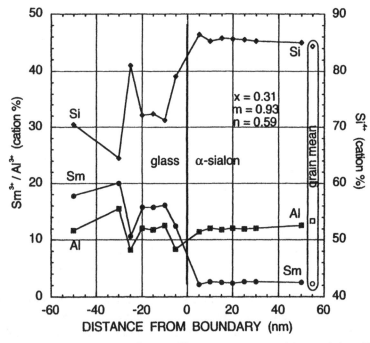

Figure 4: Elemental concentration profiles across a samarium α-sialon {10-10} prism plane growing into a pocket of residual glass.

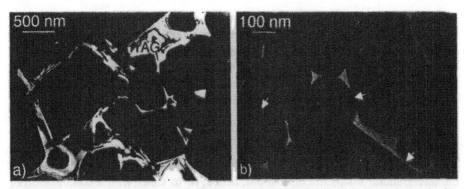

Figure 5: The microstructure of silicon carbide ceramics liquid phase sintered with the addition of 1.3 wt% Al_2O_3 and 1.7 wt% Y_2O_3 after (a) hot isostatic pressing and (b) pressureless sintering. The microstructure in (a) contains intergranular Y, Al-garnet (YAG), while the intergranular microstructure in (b) is amorphous (TEM centered dark field images). Thin grain boundary films of residual glass are arrowed in (b).

Liquid Phase Sintered Silicon Carbide Ceramics

Pressureless sintering and hot isostatic pressing (HIP) of SiC green bodies containing additions of Y_2O_3 and Al_2O_3, with an Y_2O_3/Al_2O_3 ratio corresponding to that of the Y, Al-garnet ($5Al_2O_3\cdot3Y_2O_3$), demonstrated that the sintering atmosphere as well as an applied pressure have an effect on the development of the intergranular microstructure [20]. Pressureless sintering was carried out in a SiC/Al_2O_3 protective powder bed in an argon atmosphere at 1880 °C for 2 or 4 hours. This resulted in the partitioning of Y, Al-garnet and α-Al_2O_3 from the liquid phase sintering medium, and only smaller volumes of residual glass were present in the microstructure. The glass was concentrated to thin intergranular films and to a limited number of smaller (< 20 nm) pockets. These pockets were rich in silicon, aluminum and oxygen, and contained also smaller amounts yttrium and impurities, e.g. calcium. The yttrium content of these pockets was, however far too low for the crystallization of the Y, Al-garnet.

Partitioning of the Y, Al-garnet, Figure 5 a, resulted in neighbouring pockets with the same crystallographic orientation. This suggests that the crystallization involved comparatively few nucleation sites, and that the garnet grew in a three-dimensional intergranular network. It was proposed that the formation of α-Al_2O_3 was promoted by a transport of gaseous AlO and Al_2O from the powder bed into the SiC material. The formation of AlO and Al_2O would be associated with a decomposition of the Al_2O_3 in the powder bed by carbon, which is generally present in the surrounding environment during pressureless sintering. The incorporation of Al_2O_3 from the powder bed into the intergranular microstructure was also observed for $SiC(+Y_2O_3)$ ceramics pressureless sintered under the same conditions, but without the addition of Al_2O_3.

HIP of the SiC($+5Al_2O_3\cdot3Y_2O_3$) composition at 1800 °C under a pressure of 160 MPa resulted in a completely different microstructure, see Figure 5 b. The crystallization of the liquid phase sintering medium was suppressed; Y, Al-garnet did not partition although the Y/Al ratio of analyzed residual glass pockets varied around that of the garnet. In addition, the applied high pressure during HIP caused a limited decomposition of the α-SiC resulting in the formation of intergranular graphite [20].

Silicon Nitride Based Ceramics

The intergranular microstructure in a liquid phase sintered silicon nitride based ceramic may be viewed as an oxynitride glass-ceramic. One way to controlling the intergranular microstructure would, hence, be to sinter with a combination of additives corresponding to an oxynitride glass that may be readily crystallized to a glass-ceramic after densification. A number of studies have been carried out in different metal oxide sialon systems where the intergranular microstructure also may be adjusted by the incorporation of liquid phase constituents into the solid phases, α'- and / or β'-Si_3N_4 [10, 12, 35, 39, 41, 48].

A secondary crystalline phase may also act as a reinforcement. An addition of ZrO_2 to Si_3N_4 is not only an effective sintering additive, but may also result in a composite Si_3N_4 ceramic microstructure, Figure 6, where grains of ZrO_2 act as a toughening and strengthening agent [13, 14, 47, 49]. It has been shown that the partitioning of a ZrO_2 structure from the liquid phase sintering medium makes it possible to form Si_3N_4-based ceramics with extremely small volume fractions of residual glass present only as thin grain boundary films between grains of Si_3N_4

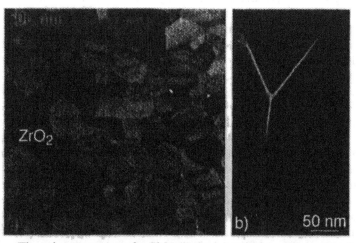

Figure 6: The microstructure of a Si_3N_4/ZrO_2 ceramic (a). The residual glass (b) is concentrated to very thin grain boundary films (TEM centered dark field image formed from diffuse scattered electrons).

and / or ZrO_2 [47, 50]. The zirconium containing oxynitride liquid phase sintering medium also promotes the development of high aspect ratio β-Si_3N_4 grains, which contributes further to the toughness of the ceramic.

The ZrO_2 may incorporate nitrogen into its structure during partitioning from the liquid phase sintering medium whereby an oxidation-prone zirconium-oxynitride solid solution is formed [49]. Work has shown that the simultaneous addition of Y_2O_3, or the use of an Y_2O_3 partially stabilized ZrO_2 starting powder, will reduce the nitrogen uptake of the $ZrO_2(+Y_2O_3)$ structure that partitions during sintering [47, 49, 51]. It was also established that the sintering atmosphere has an effect on the anion lattice [47]. HIP resulted in ZrO_2 structures with a better high temperature stability than after pressureless sintering, which infers that the nitrogen atmosphere around the Si_3N_4 powder compact during pressureless sintering results in an increased nitrogen content of the liquid and thereby of the ZrO_2 grains that form in the material. Si_3N_4/ZrO_2 ceramics have also been fabricated with a smaller addition of Al_2O_3 as sintering additive. This promotes the formation of a dilute β'-Si_3N_4 structure which will adjust the oxygen / nitrogen ratio of the liquid phase sintering medium.

The composition and structure of the ZrO_2 will determine the toughening mechanisms that may be activated [52]. Transformation toughening may take place if a transformable metastable tetragonal structure is retained in the sintered material. This structure will transform to the monoclinic structure when the mechanical constraints imposed on the ZrO_2 grains are released in a stress field. The non-transformable t' structure, which consists of tetragonal domains separated by anti-phase domain boundaries, may contribute to toughness by ferroelastic domain switching. The retention of these tetragonal structures to room temperature is realized through the incorporation of a cation with a lower valence state, e.g. Y^{3+}, into the ZrO_2 structure.

INTERGRANULAR FILMS

Even if a substantial crystallization of the intergranular regions in a liquid phase sintered silicon-based ceramic microstructure has been achieved, thin intergranular films of residual glass generally remain. Electron energy filtering and fine probe EDX analysis has demonstrated that the glassy grain boundary films in SiC and Si_3N_4 based microstructures are rich in elements originating from the metal oxide / nitride sintering additives, but depleted in carbon and nitrogen, respectively [22].

As discussed above, pressureless sintering of SiC with additions of Y_2O_3 and Al_2O_3 makes it possible to obtain a substantial crystallization of the liquid phase sintering medium. Thin films of residual glass were, however, separating the secondary Y, Al-garnet and α-Al_2O_3 structures from adjacent grains of SiC [20, 22]. Thin films of residual glass, with a varying cation content, were present also at SiC/SiC grain boundaries [20, 22]. Combined high resolution analytical and spatial information obtained from electron energy filtered images recorded around the carbon K, oxygen K and aluminum $L_{2,3}$ edges in the EEL spectrum showed that glassy grain boundary films, as well as the occasional smaller pockets of residual glass, were rich in oxygen and aluminum, but depleted in carbon, as in the HIP:ed SiC microstructure shown in Figure 7. Intensity profiles integrated along elemental images of edge-on grain boundaries, e.g. the aluminum jump ratio map in Figure 7, indicated average film thicknesses in the range 1.5 to 1.8 nm. This is consistent with results from defocus Fresnel imaging [20]. EDX showed that thin films merging into smaller yttrium and impurity cation containing glass pockets contained also these elements.

High resolution TEM of Si_3N_4-based microstructures has demonstrated that the film thickness depends upon the overall chemistry of the ceramic microstructure as well as the structure and composition of neighbouring grains. The intergranular films in β'-Si_3N_4/ZrO_2 composite ceramic microstructures may be as thin as 0.9 to 1.0 nm when the residual glass is present only at the grain boundaries [47]. Some sialon microstructures show significantly larger film thicknesses of around 2.4 nm,

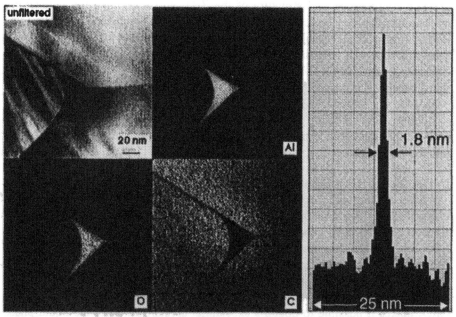

Figure 7: Elemental distribution images of a silicon carbide ceramic liquid phase sintered with the addition of 1.3 wt% Al_2O_3 and 1.7 wt% Y_2O_3. The computation of the carbon and aluminum images is shown in Figure 2. The intensity distribution in the aluminum image across the edge-on grain boundary film indicated a film thickness of 1.8 nm.

even when a substantial part of the liquid phase constituents have been incorporated into the β'- and / or α'-Si_3N_4 solid solutions, or other sialon structures, so that the volume fraction of residual glass is extremely small [22, 53].

A smaller network modifying cation tend to reduce the average thickness of the intergranular glass film [53, 54]. The replacement of Nd_2O_3 by Yb_2O_3 in the fabrication of duplex α / β-sialon ceramics resulted in a reduced film thickness, see Figure 8. The intergranular films in the microstructure containing the larger Nd^{3+} ($r(Nd^{3+}) = 0.983$ Å) were typically around 2.4 nm, while the smaller Yb^{3+} ($r(Yb^{3+}) = 0.868$ Å) resulted in residual glass films in the range 1.6 to 1.9 nm thick.

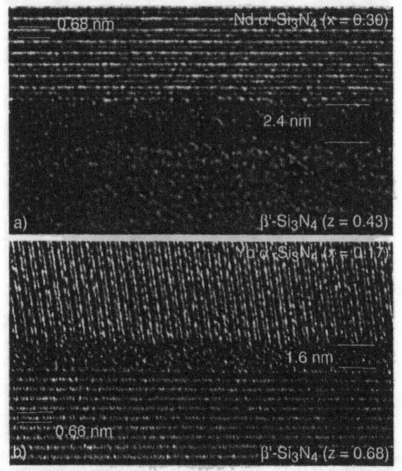

Figure 8:   Intergranular films of residual glass in duplex α / β sialons fabricated with the addition of (a) Nd_2O_3 and (b) Yb_2O_3.

The small variation in measured film thickness within a microstructure was accompanied by a variation in the local α- and β-sialon substitution levels.

The thickness of the residual glass films may, however, vary significantly within a microstructure, both between boundaries and within a particular boundary. These observations are in contrast to models suggesting that there is a constant grain boundary film thickness throughout a microstructure [55-57]. The radius of Dy^{3+} is 0.912 Å which suggests that the use of Dy_2O_3 instead of Yb_2O_3 or Nd_2O_3 in the

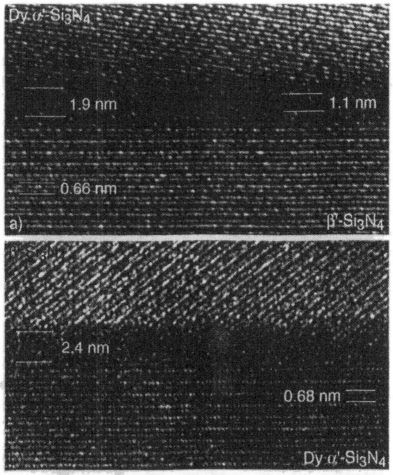

Figure 9:   Intergranular films of residual glass in a duplex α / β sialon fabricated with the addition of Dy_2O_3. The boundary in (a) has a film thickness that shows pronounced variations, while the grain boundary film in (b) has an average thickness of 2.4 nm.

fabrication of the duplex sialons discussed above would result in a film thickness in between what was observed for the smaller Yb^{3+} and larger Nd^{3+} cations. This material showed, however, a wide spread in film thickness. Figure 9 a shows an example of an α/β grain boundary where the film thickness varies between 1.1 and 1.9 nm over a distance of 20 nm along this section of the boundary. A significantly thicker film, 2.4 nm, separating an α'- and a β'-Si_3N_4 grain in the same material is shown in Figure 9 b. This microstructure showed a pronounced variation in the local substitution levels of the α'- and β'-Si_3N_4 grains, which can be expected to contribute to the observed spread in grain boundary film thickness.

Electron spectroscopic imaging and subsequent computation of elemental distribution images has been performed for a number of α- and duplex α/β-sialons fabricated with additions of Sm_2O_3, Dy_2O_3 or Yb_2O_3. The electron spectroscopic images were recorded around the nitrogen K, oxygen K, aluminum $L_{2,3}$ and rare earth element $N_{4,5}$ edges in the EEL spectrum [22]. The elemental distribution images clearly demonstrated the concentration of the α' stabilizing cation, aluminum and oxygen to the glassy grain boundary films, and that these films also have a reduced nitrogen content. These results are in accordance with elemental concentration profiles across α'/α', α'/β' and β'/β' grain boundaries, which showed an enrichment of aluminum and the rare earth element to the glassy grain boundary

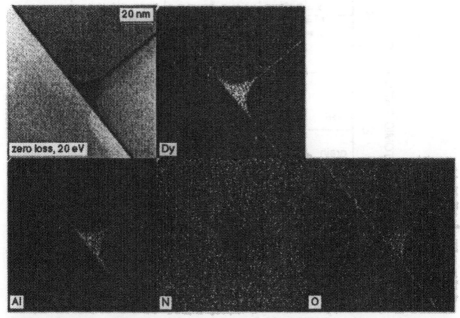

Figure 10: Elemental distribution images of a triple grain junction in a dysprosium α-sialon microstructure.

film [22]. These profiles also revealed a significant variation in the α' and β' substitution levels on a nanometer scale, which reflects the extensive solid solution of the sialon phases. An example of elemental distribution images of a sialon microstructure is shown in Figure 10, and elemental profiles from the same microstructure in Figure 11.

Different grain boundary morphologies have also been observed within a microstructure where solid solutions can not form. Si_3N_4 gas pressure sintered with 3 wt% Y_2O_3 had typically a grain boundary film thickness of 1.1 – 1.2 nm, see Figure 12 a. These films were rich in yttrium and oxygen, and contained frequently also some impurities, e.g. calcium. A rather different type of boundary was, however, also observed in this microstructure. Figure 12 b shows a high resolution image of a boundary oriented edge on in the TEM. Contrast from a set of {20-20} fringes corresponding to a lattice spacing of 0.33 nm are clearly visible in both grains. Within the resolution of the image, the grains seem to be in direct contact along a substantial part of the boundary; only some local disorder is observed.

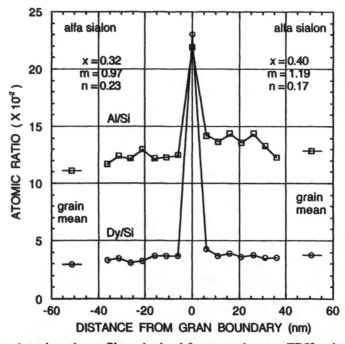

Figure 11: Atomic ratio profiles, obtained from step by step EDX point analyses, across a grain boundary film in a dysprosium α-sialon microstructure. An electron probe with a nominal FWHM of 0.7 nm was used for the acquisition of the EDX spectra.

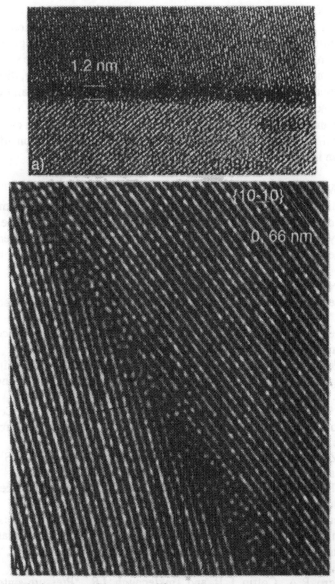

Figure 12: Grain boundary films in a silicon nitride ceramic gas pressure sintered with the addition of 3 wt% Y_2O_3. The boundary in (a) has an average film thickness of 1.2 nm. Within the resolution of the image (3.3 nm), the grains in (b) seem to be in direct contact; only some local disorder is observed between the arrows.

CONCLUDING REMARKS

Quantitative microscopy together with analytical and high resolution transmission electron microscopy provides important information which gives a deeper understanding of the development of microstructure in liquid phase sintered silicon based ceramics. The combined information from different imaging, diffraction and spectroscopic methods in the TEM is required for a qualitative and quantitative chemical and structural characterization of small volumes in the microstructure. Analytical and spatial information from regions of the size 1 to 2 nm can be obtained by electron spectroscopic imaging and subsequent computation of elemental distribution images.

ACKNOWLEDGEMENTS

Collaboration with colleagues and former students at the Swedish Ceramic Institute, the Ceramics Research Unit, University of Limerick, Ireland, the Arrhenius Laboratory, University of Stockholm and the Departments of Experimental Physics and Engineering Metals at Chalmers University of Technology is highly appreciated.

REFERENCES

[1]Lange, F.F., "Silicon nitride polyphase systems: fabrication, microstructure and properties", Int. Met. Rev., No. 1, 1980, 1-20.

[2]Lange, F.F., "Fabrication and properties of dense poly-phase silicon nitride", J. Am. Ceram. Soc. Bull., 1983, 62, 1369-74.

[3]Ziegler, G., Heinrich, J. and Wötting, G., "Review: relationships between processing, microstructure and properties of dense and reaction-bonded silicon nitride", J. Mater. Sci., 1987, 22, 3041-3086.

[4]Jack, K.H., "Silicon nitride, sialons, and related ceramics". In Ceramics and Civilization, Vol. III, High-Technology Ceramics. American Ceramic Society, Columbus, OH, 1986, pp. 259-88.

[5]Jou, Z.C., Virkar, V. and Cutler, R.A., "High temperature creep of SiC densified using transient liquid phase", J. Mater. Res., 1991, 6, 1945-1949.

[6]Raj, R., "Fundamental research in structural ceramics for service near 2000 °C", J. Am. Ceram. Soc., 1993, 76, 2147-2174.

[7]Cutler, R.A. and Jackson, T.B., "Liquid phase sintered silicon carbide". In Proceedings of the 3rd International Symposium on Ceramic Materials and Components for Engines, ed. V.J. Tennery. American Ceramic Society, Westerville, OH, 1989, pp. 309-318.

[8]Alliegro, R.A., Coffin, L.B. and Tinklepaugh, J.R., "Pressure-sintered silicon carbide", J. Am. Ceram. Soc., 1956, 39, 386.

[9]Tanaka, H., "Sintering of silicon carbide". In Silicon Carbide Ceramics, eds. S. Somiya and Y. Inomata. Elsevier Applied Science, London, 1991, pp. 213-238.

[10]Lewis, M.H. and Lumby, R.J., "Nitrogen ceramics: liquid phase sintering", Powder Metallurgy, 1983, 26, 73-81.

[11]Cordrey, L., Niesz, D.E. and Shanefield, J., "Sintering of silicon carbide with rare-earth oxide additions". In Sintering of Advanced Ceramics, eds. C.A. Handwerker, J.E. Blendell and W. Kaysser. The American Ceramic Society, Inc. Westerville, Ohio, 1990, pp. 618-636.

[12]Lewis, M.H., Leng-Ward, G. and Jasper, C., "Sintering additive chemistry in controlling microstructure and properties of nitrogen ceramics", Ceram. Powder Sci., 1988, 2, 1019-1033.

[13]Falk, L.K.L. and Rundgren, K., "Microstructure and short term oxidation of hot-pressed Si_3N_4/ZrO_2 (+Y_2O_3) ceramics", J. Am. Ceram. Soc., 1992, 75 [1], 28-35.

[14]Knutson-Wedel, M., Falk, L.K.L. and Ekström, T., "Characterization of Si_3N_4 ceramics formed with different oxide additives", J. Hard Materials, 1992, 3 [3-4], 435-445.

[15]Becher, P.F., "Microstructural design of toughened ceramics", J.Am. Ceram. Soc., 1991, 74, 255-269.

[16]Karunaratne, B.S.B. and Lewis, M.H., "High-temperature fracture and diffusional deformation mechanisms in Si-Al-O-N ceramics", J. Mater. Sci., 1980, 15, 449-462.

[17]Knutson - Wedel, E.M., Falk, L.K.L., Björklund, H. and Ekström, T., "Si_3N_4 ceramics formed by HIP using different oxide additions - relations between microstructure and properties", J. Mater. Sci., 1991, 26, 5575-5584.

[18]Tuersley, I.P., Leng-Ward, G. and Lewis, M.H., "High-temperature Si_3N_4 ceramics". In Engineering with Ceramics, ed. R. Morrell. British Ceramic Society, 1990, pp. 231-246.

[19]Hampshire, S. and Jack, K.H., "The kinetics of densification and phase transformation of nitrogen ceramics". In Special Ceramics 7, eds. D. Taylor and P. Popper. British Ceramic Research Association, Stoke-on-Trent, 1981, pp. 37-49.

[20]Falk, L.K.L., "Microstructural development during liquid phase sintering of silicon carbide ceramics", J. Eur. Ceram. Soc., 1997, 17, 983-994.

[21]H. Björklund, J. Wasén and L.K.L. Falk, "Quantitative Microscopy of b-Si3N4 Ceramics", J. Am. Ceram. Soc. 80 (1997) 3061-69.

[22]L.K.L. Falk, "Electron spectroscopic imaging and fine probe EDX analysis of liquid phase sintered ceramics", J. Eur. Ceram. Soc., 18 (1998) 2263-2279.

[23]Krivanek, O.L., Shaw, T.M. and Thomas, G., "Imaging of thin intergranular phases by high-resolution electron microscopy", J. Appl. Phys., 1979, 50, 4223-4227.

[24]Clarke, D.R., "On the detection of thin intergranular films by electron microscopy", Ultramicroscopy, 1997, 4, 33-44.

[25]Ness, J.N., Stobbs, W.M. and Page, T.F., "A TEM Fresnel diffraction-based method for characterizing thin grain-boundary and interfacial films", Phil.Mag. A, 1986, 54, 679-702.

[26]Cinibulk, M.K., Kleebe, H.-J. and Rühle, M., "Quantitative Comparison of TEM Techniques for Determining Amorphous Intergranular Film Thickness", J. Am. Ceram. Soc., 1993, 76, 426-432.

[27]Gubbens, A.J. and Krivanek, O.L., "Applications of a post-column imaging filter in biology and materials science", Ultramicroscopy, 1993, 51, 146-159.

[28]Berger, A., Mayer, J. and Kohl, H., "Detection limits in elemental distribution images produced by energy filtering TEM: case study of grain boundaries in Si3N4", Ultramicroscopy, 1994, 55, 101-112.

[29]Hofer, F., Grogger, W., Kothleitner, G. and Warbichler, P., "Quantitative analysis of EFTEM elemental distribution images", Ultramicroscopy, 1997, 67, 83-103.

[30]Lewis, M.H., Powell, B.D., Drew, P., Lumby, R.J., North, B. and Taylor, A.J., "The formation of Single-phase Si-Al-O-N- ceramics", J. Mater. Sci. 1977, 12, 61-74.

[31]Wötting, G., Kanka, G. and Ziegler, G., "Microstructural development, microstructural characterization and relation to mechanical properties of dense silicon nitride". In Non-oxide Technical and Engineering Ceramics, ed. S. Hampshire. Elsevier Applied Science, London, 1986, pp. 83-96.

[32]P. Obenaus aand M. Herrmann, "Qualitatively characterising columnar crystals in silicon nitride ceramics", Pract. Met. 1990, 27, 503-513.

[33]Björklund, H. and Falk, L.K.L., "Grain morphology and intergranular microstructure of whisker reinforced Si_3N_4 ceramics", J. Eur. Ceram. Soc., 1997, 17, 13-24.

[34]Björklund, H., Falk, L.K.L., Rundgren, K. and Wasén, J., "β-Si_3N_4 grain growth, Part I: Effect of metal oxide additives", J. Eur. Ceram. Soc., 1997, 17, 1285-1299.

[35]Falk, L.K.L. and Dunlop, G.L., "Crystallisation of the glassy phase in a Si_3N_4 Material by post-sintering heat treatments", J. Mater. Sci., 1987, 22, 4369-4376.

[36]Hampshire, S., Park, H.K., Thompson, D.P. and Jack, K.H., "α-Sialon Ceramics", Nature (London), 1978, 274, 880-82.

[37]Jack, K.H., "The characterization of a-sialons and the a-b relationships in sialons and silicon nitrides". In Progress in Nitrogen Ceramics, ed. F.L. Riley. Martinus Nijhoff Publishers, The Hague, 1983, pp. 45-60.

[38]Ekström, T. and Nygren, M., "Sialon ceramics", J. Am. Ceram. Soc., 1992, 75, 259-276.

[39]Ekström, T., Falk, L.K.L. and Shen, Z.-J, "Duplex α-β sialon ceramics stabilized by dysprosium and samarium", J. Am. Ceram. Soc., 1997, 80, 301-312.

[40]Falk, L.K.L., Shen, Z.-J. and Ekström, T., "Microstructural stability of duplex α-β sialon ceramics", J. Eur. Ceram. Soc., 1997, 17, 1099-1112.

[41]Thompson, D.P., "New grain-boundary phases for nitrogen ceramics". In Silicon Nitride Ceramics: Scientific and Technological Advances, eds. I-W. Chen, P.F. Becher, M. Mitomo, G. Petzow and T.-S. Yen. Materials Research Society, Pittsburgh, Pennsylvania, 1993, pp. 79-92.

[42]Clarke, D.R. and Lange, F.F., "Oxidation of Si_3N_4 alloys: Relation to phase equilibria in the system Si_3N_4-SiO_2-MgO", J. Am. Ceram. Soc., 1980, 63, 586-593.

[43]Raj, R. and Lange, F.F., "Crystallisation of small quantities of glass (or liquid) segregated in grain boundaries", Acta Metallurgica, 1981, 29, 1993-2000.

[44]Kessler, H., Kleebe, H.-J., Cannon, R.W. and Pompe, W., "Influence of internal stresses on crystallization of intergranular phases in ceramics", Acta metall. mater., 1992, 40, 2233-2245.

[45]Björklund, H. and Falk, L.K.L., "β-Si_3N_4 grain growth, Part II: Intergranular glass chemistry", J. Eur. Ceram. Soc., 1997, 17, 1301-1308.

[46]Pullum, O.J. and Lewis, M.H., "The effect of process amtosphere on the intergranular phase in silicon nitride ceramics", J. Eur. Ceram. Soc., 1996, 16, 1271-1275.

[47]Knutson-Wedel, M., "The microstructure of metal oxide additive silicon nitride ceramics" (ISBN 91 – 7197 – 280 – 3). Ph.D. thesis, Chalmers University of Technology, Göteborg, Sweden, 1996 .

[48]Cheng, Y.B. and Thompson, D.P., "Preparation and grain boundary devitrification of samarium α-sialon ceramics", J. Eur. Ceram. Soc., 1994, 14, 13-21.

[49]Lange, F.F., Falk, L.K.L. and Davies, B.I., "Structural Ceramic Composites Based on Si_3N_4-ZrO_2(+Y_2O_3) Compositions", J. Mater. Res., 1987, 2, 66-76.

[50]Falk, L.K.L., "Microstructural Development of Si_3N_4 Ceramics Formed with Additions of ZrO_2", Materials Forum, 1993, 17, 83-93.

[51]Cheng , Y.B. and Thompson, D.P., "Role of anion vacancies in nitrogen-stabilized zirconia", J. Am. Ceram. Soc., 1993, 76, 683.

[52]Green, D.J., Hannink, R.H.J. and Swain, M.V., Transformation toughening of Ceramics (CRC Press Inc., Florida, USA, 1989).

[53]Falk, L.K.L., "Imaging and Analysis of Sialon Interfaces". In Nitrides and Oxynitrides, Materials Science Forum Vols. 325-326, eds. S. Hampshire and M.J. Pomeroy. Trans Tech Publications, Switzerland, 2000, pp. 231-236.

[54]Wang, C.-M., Pan, X., Hoffmann, M.J., Cannon, R.M. and Rühle, M., J.Am.Ceram.Soc., 1996, 79, 788-792. .

[55]Kleebe, H.-J., Cinibulk, M.K., Cannon, R.M. and Rühle, M.R., "Statistical analysis of the intergranular film thickness in silicon nitride ceramics", J. Am. Ceram. Soc., 1993, 76, 1969-1977.

[56]Clarke, D.R., "On the equilibrium thickness of intergranular glass phases in ceramic materials", J. Am. Ceram. Soc., 1987, 70, 15-22.

[57]Clarke, D.R., Shaw, T.M., Philipse, A.P. and Horn, R.G., "Possible electrical double-layer contribution to the equilibrium thickness of intergranular glass films in polycrystalline ceramics", J. Am. Ceram. Soc., 1993, 76, 1201-1204.

GRAIN-BOUNDARY RELAXATION PROCESSES IN SILICON-BASED
CERAMICS STUDIED BY MECHANICAL SPECTROSCOPY

Giuseppe Pezzotti and Ken'ichi Ota
Ceramic Physics Laboratory, Department of Materials,
Kyoto Institute of Technology,
Sakyo-ku, Matsugasaki, 606-8585 Kyoto, Japan

ABSTRACT
   Grain boundaries in polycrystalline ceramics are characterized by peculiar
atomic structures. Two kinds of pattern may be recognized: (i) a glassy phase fills
the grain boundaries and multiple-grain pockets; and, (ii) the grain boundaries
show a "clean" structure without secondary phases. Glassy phases usually form a
film of nanometer scale thickness, which encompasses the grain structure.
Mechanical spectroscopy conducted at high temperatures and low frequencies
may provide a unique characterization tool in selectively envisaging the inherent
rheological properties of intergranular glassy phases. According to the peak-shift
method, the viscosity magnitude of the intergranular glass and its activation
energy can be quantitatively estimated. It is newly shown here that also in the case
of ceramic polycrystals with glass-free boundaries, a quantitative knowledge of
elementary phenomena behind the deformation behavior can be achieved.

INTRODUCTION
   The accumulation of knowledge, particularly from high-resolution electron
microscopy investigations, concerning the atomic structure of ceramic grain
boundaries makes increasingly realistic a "grain-boundary engineering" approach
to the development of new ceramic materials. It can be interesting, at this stage of
development and in the light shed by the new structural understanding of ceramic
grain boundaries, to revisit and critically re-discuss the two earliest conceptions of
the structure of grain boundary: (i) the "amorphous cement" theory; and (ii) the
"transitional lattice" theory. In a context of advanced ceramic science, it is
surprising to see how, although both theories were developed when experimental
evidence as to the nature of grain boundaries could only be obtained through
indirect methods (*e.g.*, macroscopic properties measurements), they preserve
fundamental validity. More importantly, we shall show here experimental
evidence that in polycrystalline ceramics these two theories are both applicable
and compatible with each other. An additional goal of this paper is to describe
some of the most recent developments in the characterization of the

micromechanical behavior of ceramic grain boundaries. In particular, we shall focus our description on mechanical spectroscopy techniques, which aim at characterizing the resistance to sliding of grain boundaries from the rate of the observed mechanical energy dissipation in the polycrystal. In mechanical spectroscopy, an oscillation of given frequency is sent as a probe through the tested specimen and the rate of internal energy dissipation is measured by detecting the decrease in oscillation amplitude or the phase lag between stress and strain. The existence of one or more maxima in the internal energy dissipation (either as a function of temperature or frequency) gives rise to the so called "mechanical spectrum", whose interpretation may enable one understanding micromechanical phenomena occurring at grain boundaries. As it might be expected, the simplest cases of high-purity materials are those which have received the most experimental study, approaching a satisfactory quantitative treatment in terms of an explicit atomic model. It is for this reason that a disproportionate fraction of research has been so far devoted to grain boundaries of "model" materials. However, we understand that the ideas developed in connection with model grain boundaries may be extended to internal boundaries of a more general class.

BACKGROUND ON GRAIN-BOUNDARY STRUCTURE AND PROPERTIES
Historical overview on grain-boundary structures

The first theory for grain boundaries, which was developed in detail and which, with minor modifications, has been widely accepted for many years was the "amorphous cement" theory. Radavich[1], Rosenhain[2], and Rosenhain and Ewen[3] proposed that a "glassy cement" may be systematically present at grain boundaries of polycrystalline metals. To support this hypothesis two experimental evidences were invoked; first, Beilby[4] proved the formation of a supercooled liquid on surfaces of polished metal crystals, as a result of localized melting/flow processes followed by solidification. This supercooled liquid (or amorphous material) was considered to have the characteristics of a vitreous material. Accordingly, Bengough[5] first identified the grain boundary cement with Beilby's amorphous metal; several years later, Chalmers[6] (for high-purity tin) and Chaudron et al.[7] (for aluminum) experimentally proved that the apparent melting point of grain boundaries was actually lower than that of the bulk grains, independent of the difference in orientation of the grains.

In 1929, Hargreaves and Hills[8] proposed the theory that the grain boundary is a transition from the lattice of one grain to that of the other (i.e., the transitional lattice theory). They argued that in an aggregate of grains the forces which caused the atoms to assume the regular arrangement of a crystal lattice must still operate at the boundaries, causing the atoms at the boundaries to take up positions dictated by forces exerted by the atoms within the crystals. Clearly, if both surfaces are of arbitrary orientation, many atoms in the neighborhood of the interface will be much too close or much too far away from an ideal equilibrium distance. The atoms will thus move to new positions in the immediate neighborhood of the boundary, the radical readjustments of atoms nearest the boundary forcing smaller readjustments also in atoms up to a few lattice distances away from the boundary. The grain boundary will consist of a transition between the two orientations, which can be described by a radical readjustment of atomic bonds across the boundary plus a primarily elastic distorsion through a thin transition layer on either side of the boundary. In the transitional lattice concept, the atomic arrangement will be the same for any two boundaries where the orientation of the grains on either side and the direction of the boundary are the same. On the transition hypothesis, the degree of disorder at the boundary will depend upon the change of orientation between the grains, and it would be expected that the properties of a boundary would depend upon the relative orientation of the grains. If, on the other hand, the boundary structure is amorphous, then the properties of this layer should not change with the orientation of the grains. Translating these concepts in terms of a mechanical spectrum, one should expect to observe a single, intense maximum for amorphous boundaries with an uniform structure throughout the polycrystal. On the other hand, several maxima should be observed in polycrystals with inhomogeneous grain-boundary structure, each one corresponding to different grain-boundary structures of statistical relevance within the polycrystal.

Micromechanical assessments of grain boundaries

Zener[9] has pointed out that various aspects of the mechanical behavior of polycrystalline bodies may be explained on the assumption that the resistance to slipping of grain boundaries obeys the laws commonly associated with amorphous materials rather than the laws associated with crystalline materials. The resistance to slip at grain boundaries increases rapidly with increase in deformation rate and with decrease in temperature. The Zener's statement represents the first step towards the development of a grain-boundary micromechanics field. The hypothesis that grain boundaries behave in a viscous manner and allow the relaxation of a shear stress across them has been confirmed through mechanical spectroscopy studies. The first micromechanical evidence for grain-boundary viscosity, obtained by means of mechanical spectroscopy, was provided by Kê[10] for polycrystalline aluminum. Assuming an effective thickness for the grain boundary of 0.4 nm (or one atomic distance), Kê obtained a grain-boundary viscosity value of 1.4 x 10^{-2} Pa x s at around 290 °C, which is in agreement with the viscosity value experimentally determined by Polack and Sergueiev[11] for liquid aluminum. However, Kê's correlation has long been regarded as fortuitous and a popular theory of grain-boundary slip (theory of the "coherency islands") given by Mott[12] also led many researchers in the metal field to this same conclusion. Michaud[13], for example, concluded from his own extensive experiments on the effect of grain boundaries on plastic flow in pure aluminum that the results generally are inconsistent with the model of an amorphous grain-boundary cement. Michaud regarded the boundaries purely as surfaces of separation between two grains, with no distinctive mechanical properties of their own. After long debates, Kê finally stated[14]: "The viscous behavior of grain boundaries gives no definite information regarding the nature or the structure of the boundary region between adjacent crystals." Actually, besides the importance of the pioneering work made by Kê, the extent to which mechanical spectroscopy was developed at that time did not allow quantitative characterizations of grain-boundary properties. Nowadays, at the Ceramic Physics Laboratory of Kyoto Institute of Technology, we are able to automatize our experiments, rising the measurement temperature up to 3300 K and expanding the frequency range over several orders of magnitude (from mHz to several tens Hz). This expanded parameter range, together with the availability of high-resolution electron microscopy characterizations, makes it possible a quantitative evaluation of grain-boundary-related phenomena by means of mechanical spectroscopy. By the light of these improved assessments, we can presently discuss the grain-boundary structure of polycrystalline ceramics and confidently state that whatever the structure of such boundaries, either glassy-like or directly bonded, they will undergo sliding, if the required time or temperature is provided. In particular, the concept of viscous behavior of grain boundaries is consistent with both principal theories regarding the nature of grain boundaries, *i.e.*, the amorphous cement and the transitional lattice theory. More importantly, from the sliding behavior of the grain boundaries we can better understand details of grain-boundary structure with some statistical validity.

Mechanical spectra as a function of temperature/frequency

A polycrystalline solid may be described as showing a "discrete spectrum" of relaxation processes (henceforth simply referred to as the "mechanical spectrum"). Spectra can be plotted as a function of temperature (at a fixed oscillation frequency) or as a function of frequency (at a given temperature). The peaks contained in the spectrum are usually characterized by a broadened configuration, which may arise *e.g.* from a sub-spectrum of relaxation times related to the geometrical and physical parameters involved in the relaxation phenomenon.

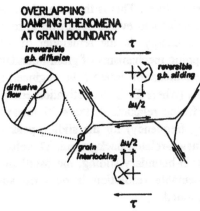

OVERLAPPING
DAMPING PHENOMENA
AT GRAIN BOUNDARY

Fig.1: Slip along and diffusional flow across grain boundaries.

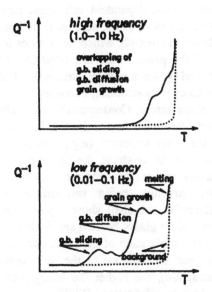

Fig.2: Schematic of mechanical spectra, $Q^{-1}(T)$, at different frequencies.

However, the observed mechanical spectrum is a result of the superposition of several thermally activated mechanisms, which may have different activation energies. In polycrystalline ceramics, the mechanical spectrum can be the result of the superposition of grain-boundary sliding, grain-boundary diffusion and migration (ultimately leading to grain growth) and bulk diffusion (ultimately leading to melting). Each of these features of the spectrum obeys a different activation energy value, therefore the spectrum significantly changes its morphology as a function of oscillation frequency. It is simply necessary to realize that the change in the resolution of different spectrum features is a measure of the difference in their corresponding activation energies. This result follows from the fact that the larger the activation energy value, the smaller the shift of the internal friction peak (for a given change in frequency). A typical feature revealed in the mechanical spectrum of polycrystalline ceramics is an overlapping process of grain-boundary sliding and intergranular diffusion/migration; the former involving a reversible movement of matter along the boundaries, while the latter leads to irreversible flow of matter along and across the boundaries (cf. schematic in Fig. 1). Sliding gives rise to a maximum in the mechanical spectrum, while diffusion leads to a background curve, which rises with an exponential-like trend (until incipient grain growth occurs). Figure 2 shows two schematics of the relaxation spectrum of polycrystalline ceramics as recorded both at high and low frequencies. It can be noticed that the mechanical spectrum unfolds, resolving more features. This is the case when the measurement is pushed towards lower frequencies and events occurring at more elevated temperatures have higher activation energy. In the remainder of this paper, we focus on characterizing the overlapping mechanisms of grain-boundary sliding and diffusion, these being the phenomena of interest in engineering assessments. It should be noted that, if $E_s < E_d$ (where E_s is the activation energy for sliding relaxation and E_d is that for diffusion relaxation), the shift towards lower temperatures observed (upon lowering frequency) for the sliding peak is relatively more marked than that of the diffusion-related background. Therefore, for resolving the spectroscopic details of grain-boundary sliding, the oscillation frequency should be lowered until an acceptable resolution is obtained and sliding/diffusion phenomena are properly separated.

MECHANICAL SPECTRA OF MODEL Si-BASED POLYCRYSTALS

When considering the relaxation peak arising from grain-boundary sliding, three quantities are of interest: (i) the temperature-dependent relaxation time, which determines the peak position on a temperature scale, and its shift upon changing oscillation frequency; (ii) the relaxation strength, which determines the peak intensity; and, (iii) the morphology of the peak, which is related to both activation energy for intergranular flow and grain-boundary structure. With increasing temperature, a grain boundary, whatever its structure, becomes viscous and cannot sustain an externally applied shear stress. Consequently, local slip occurs until the shear stress along grain boundaries has been released. Provided that no permanent damage occurs in the microstructure (*e.g.*, wedge-like microcracking, cavitation, etc.), upon releasing external stress, the elastic contact stresses at the positions where grains interlock will produce the recovery of the original morphology, the polycrystalline structure being restored to its undeformed state (anelastic relaxation). Hereafter, we shall show experimental evidence that grain-boundary sliding occurs both in glassy bonded and in directly bonded grain boundaries. A phenomenological implication of the grain-boundary sliding mechanism is that the higher the peak temperature in the mechanical spectrum (at given frequency and microstructure), the higher the intergranular glass viscosity. Therefore, by comparing the internal friction, $Q^{-1}(T)$, curves of different polycrystals with respect to their sliding relaxation peak (recorded at the same oscillation frequency), one can acquire comparative information about the softening temperature of grain boundaries. Similarly, if different grain-boundary structures are present within the same polycrystal, different grain-boundary peaks should be observed whose position in the temperature scale depends on their sliding resistance, the higher the peak-maximum temperature the higher the

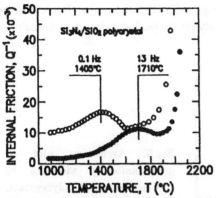

Fig.3: Mechanical spectrum of model Si₃N₄ polycrystal

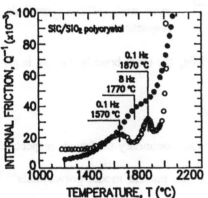

Fig.4: Mechanical spectrum of model SiC polycrystal.

resistance. Figures 3 and 4 compare the mechanical spectrum recorded at different frequencies for two model polycrystals of Si_3N_4 and SiC, both containing a small fraction (2.5 vol.%) of SiO_2. At high frequencies, both materials similarly show a peak arising from grain-boundary sliding. A comparison between peaks arising from SiO_2-glass-wetted grain boundaries shows that intergranular SiO_2 in SiC softens at higher temperature as compared to the intergranular SiO_2 segregated to Si_3N_4. It has been shown[15-17] that this difference arises from a peculiar structure of the intergranular SiO_2 network, including aliovalent anions of either N or C (in Si_3N_4 and SiC, respectively). In addition, it can be clearly seen that the peak in Si_3N_4 nearly remains morphologically unchanged even when measurements are pushed towards low oscillation frequencies. This suggests the presence of only one type of grain boundaries (i.e., boundaries wetted by SiO_2 glass, as previously confirmed by electron microscopy observation[15]). On the other hand, lowering the oscillation frequency in polycrystalline SiC makes it appearing a sharp peak, in addition to the peak arising from glass-wetted boundaries (this latter peak was already reported in Ref. 16). This additional sharp peak can possibly represent the sliding behavior of a non-negligible fraction of directly bonded boundaries. These boundaries are clearly more resistant to sliding than those wetted by SiO_2 glass, the sliding peak being located at higher values in a temperature scale. The presence of both wetted and unwetted boundaries in the SiO_2-containing SiC polycrystal is supported by systematic transmission electron microscopy analysis, which has been presented in a previous report[16]. It is important to emphasize here that a completely new piece of information has become available by recording mechanical spectra at low oscillation frequencies. It has been so far established that in SiO_2 wetted boundaries, the viscosity (at the peak-maximum temperature,

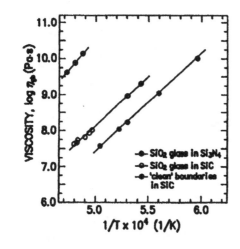

Fig.5: Arrhenius plot of grain-boundary viscosity in Si-based ceramic polycrystals. The slopes correspond to activation energies of 410, 415, and 550 kJ/mol for close, open and half-open symbols, respectively.

T_p) of the intergranular glassy phase, η_{gb}, filling the grain-boundary films, is given by[18]:

$$\eta_{gb} = G\delta/2\pi f\alpha(1-v)d \qquad (1)$$

where d and δ are the grain size and the grain-boundary thickness, respectively

(typically 1 μm and 1 nm, respectively, in the present model polycrystals[15,16]). Note that an increase in the selected oscillation frequency, f, will reduce the intergranular viscosity value at which the peak maximum manifests, thus shifting the peak towards higher temperatures. α is an adimensional constant (usually referred to as relaxation factor), which relates to the peak intensity (relaxation strength) and takes into consideration both geometry of grain-to-grain contacts and specimen geometry. α, can be calculated according to the following equation[15]:

$$\alpha = [(G/G_r)-1]/(1-v) \qquad (2)$$

where the relaxation ratio, G/G_r, between the shear modulus at room temperature

Silicon-Based Structural Ceramics

and at $T=T_p$, is an experimentally accessible parameter. Using eqs(1) and (2), an Arrhenius plot of viscosity for SiO_2-based intergranular glasses can be obtained as shown in Fig. 5. As seen, aliovalent C anions can greatly enhance the viscosity of the glass network, above the effect of aliovalent N, as already discussed in previous reports[15-17]. Sliding of directly (covalently) bonded ceramic grain-boundaries has not yet been reported in the literature and we report here for the first time about the existence of this peak. The present study also shows that, at relatively high oscillation frequencies, this peak is "washed out" into a diffusional background and thus it is not spectroscopically resolvable. A quantitative calculation using eqs.(3) and (4) can be attempted with considering a "film thickness" of the order 0.1 nm (*i.e.*, the interatomic distance). The result of this calculation is also shown in Fig. 5 and suggests an inherent viscosity for these boundaries more than two orders of magnitude higher than that of SiO_2-glass-wetted grain boundaries.

CONCLUSION

Mechanical spectroscopy measurements were systematically performed to analyze the intrinsic rheological behavior of grain boundaries in polycrystalline Si_3N_4 and SiC ceramics. Experimental evidence is provided that: (i) solution of C anions into the grain-boundary network is more effective than that of N anions in enhancing the viscosity of the intergranular SiO_2 glass; (ii) the intergranular structure of polycrystalline SiC is intrinsically different from that of Si_3N_4, due to the presence of a large fraction of directly bonded (*i.e.*, glassy film-free) grain boundaries. A sharp internal friction peak, only collected at very high temperature and low frequencies, suggests that such boundaries possess statistical relevance as an independent type of boundaries within the SiC polycrystal and that they are very resistant to sliding. From a general viewpoint, we have shown that the mechanical spectrum of polycrystalline ceramics can be unfolded by lowering the oscillation frequency, thus providing unique insight into the elementary mechanisms behind the deformation behavior of these technologically important materials.

ACKNOWLEDGEMENT: This research was supported by the NEDO International Joint Research Grant (for the fiscal year 2002) entitled "Quantum-Structure and Micromechanics of Ceramic Grain Boundaries".

REFERENCES
[1]J. F. Radavich, "The study of Grain Boundaries with the Electron Microscope," Journal of Metals; *Transactions, American Institute of Mining and Metallurgical*

Engineering, **185,** 395-402 (1949).

[2]W. Rosenhain, "The Present Position of the Amorphous Theory," *Metallurgist,* **1,** 2-7 (1925).

[3]W. Rosenhain and D. Ewen, "The Intercrystalline Cohesion of Metals," *Journal, The Institute of Metals,* **10,** 119-24 (1913).

[4]G. T. Beilby, "The Hard and Soft States in Metals," *Proceedings, Royal Society,* **72,** 218-24 (1903).

[5]G. D. Bengough, "A Study of the Properties of Alloys at High Temperatures," *Journal, The Institute of Metals,* **7,** 123-28 (1912).

[6]B. Chalmers, "Crystal Boundaries in Tin," *Proceedings, Royal Society, Series A,* **175,** 100-06 (1940).

[7]G. Chaudron, P. Lacombe, N. Yannaquis, "On the Behavior of Grain Junctions during the Melting of Pure Aluminum," *Comptes rendus,* **226,** 1372-78 (1948).

[8]F. Hargreaves and R. J. Hills, "Work-Softening and a Theory of Crystalline Cohesion," *Journal, The Institute of Metals,* **41,** 257-62 (1929).

[9]C. Zener, "Anelasticity of Metals," *Transactions, American Institute of Mining and Metallurgical Engineers,* **167,** 155-58 (1946).

[10]T. S. Kê, "Experimental Evidence of the Viscous Behavior of Grain Boundaries in Metals (Aluminum and Magnesium)," *Physical Review,* **71,** 533-38 (1947).

[11]E. V. Polack and S. V. Sergueiev, "Determination of the Viscosity of Molten Aluminum and its Alloys," *Comptes rendus,* **30,** 137-41 (1941).

[12]N. F. Mott, "Slip at Grain Boundaries and Grain Growth in Metals," *Proceedings, Physical Society,* **60,** 391-98 (1948).

[13]R. Michaud, "Contributions to the Study of of the Mutual Reactions of Crystals in the Deformation of Polycrystalline Metals," Scientific and Technical Publication, Air Ministry (France), No. 240, 1950.

[14]T. S. Kê, "Viscous Slip Along Grain Boundaries and Diffusion of Zinc in Alpha brass," *Journal of Applied Physics,* **19,** 285-89 (1948).

[15]G. Pezzotti, K. Ota, H.-J. Kleebe, "Grain-Boundary Relaxation in High-Purity Silicon Nitride," *Journal of the American Ceramic Society,* **79** [9] 2237-46 (1996).

[16]G. Pezzotti, H.-J. Kleebe, and K. Ota, "Grain-Boundary Viscosity of Polycrystalline Silicon Carbides," *Journal of the American Ceramic Society,* **81** [12] 3293-99 (1998).

[17]G. Pezzotti, "Local Viscosity of Si-O-C-N Nanoscale Amorphous Phases at Ceramic Grain Boundaries," *Journal of the American Ceramic Society,* **84** [1] 2225-28 (2001).

[18]D. R. Mosher and R. Raj, "Use of the Internal Friction Technique to Measure Rates of Grain Boundary Sliding," *Acta Metallurgica,* **22** [12] 1469-74 (1974).

HIGH TEMPERATURE STIFFNESS AND DAMPING TO QUANTITAVELY ASSESS THE AMORPHOUS INTERGRANULAR PHASE IN SINTERED SILICON NITRIDE AND CARBIDE

G. Roebben, C. Sarbu, O. Van der Biest
Dept of Materials Science and Engineering, K.U.Leuven
Kasteelpark Arenberg 44
B-3001 Heverlee
Belgium

This paper shows how mechanical spectroscopy (i.e. the measurement of stiffness and damping properties as a function of e.g. temperature) can be used to quantitatively assess the intergranular glass phases in liquid-phase-sintered silicon nitrides and carbides. A recently proposed relation is presented which links the change of stiffness as a function of temperature with the volume fraction of intergranular amorphous phase. Current issues in the field of high-temperature mechanical spectroscopy of silicon-based ceramics such as high-strain-amplitude uniaxial tests, and the relation between damping and fatigue and fracture toughness, are discussed.

LIQUID PHASE SINTERED SILICON-BASED CERAMICS

To obtain fully dense silicon-based ceramics (silicon nitride or silicon carbide) one often uses oxide additives. In this so-called liquid phase sintering (LPS) process the additives form, together with the SiO_2-layer on the surface of the Si_3N_4 and SiC powder particles, a eutectic, low-melting point phase. After sintering, remnants of the liquid phase remain at the grain boundaries. This intergranular fraction is at least partially amorphous and has a profound influence on the properties of the ceramic, especially above the glass transition temperature, T_g, of the intergranular glass.

The overall picture of the liquid phase sintered microstructure is that of a 80 to 99 vol% phase of small ($\pm$ 1 μm) ceramic grains embedded in a wetting phase, the residue of the oxide sintering additives. The volume fraction of intergranular phase (IGP) differs from the total amount of sintering additives, due to solid solution effects and material loss during processing. In addition, the IGP incorporates the SiO_2 that covers the non-oxide starting powder if it has been exposed to air. The largest fraction of the IGP is located in small volumes at junctions of three or more grains. Ideally, one should distinguish between triple-junction volumes, which are 'channels' along the lines where three grains meet, and multi-grain-junction volumes. The triple-junction channels interconnect the latter multi-grain-junction pockets. The volume fraction of intergranular pockets and triple junctions can be estimated from analysis of SEM-images of polished and plasma-etched surfaces. This procedure does not distinguish the crystalline and amorphous fractions of the IGP.

A smaller but equally important amount of IGP can be found as thin amorphous films between neighbouring grains. The model of Clarke [1] supports the observation of an equilibrium but material-dependent film width, typically around 1 nm. The model assumes that, if the IGP wets the grains, a force balance establishes between an attractive van der Waals dispersion force and a repulsive steric force acting across the grain boundary. At small film widths, the steric, repulsive force becomes very large, which is why complete elimination of the IGP from the grain boundary, e.g. under action of mechanical loads, is very difficult. A major issue in the modelling of the high temperature behaviour of LPS ceramics is the lack of information on the viscosity of the intergranular glass film. This viscosity depends on the glass composition, but direct analysis is very difficult, for reasons of spatial resolution [2].

CREEP OF LIQUID PHASE SINTERED CERAMICS

At temperatures typically in excess of 1000°C, the deformation behaviour of LPS-ceramics becomes increasingly viscoplastic. Viscous flow of glassy IGPs contributes to the creep deformation. Wilkinson and his co-workers [3,4] have demonstrated how the stress-induced redistribution of amorphous matter can dominate primary creep. This deformation results in a bimodal distribution of grain boundary film widths, with a smaller-than-average value for films on grain boundaries across which a compression force exists, and a larger-than-average value for the films under tension [5]. After unloading the material will slowly re-establish the equilibrium film thickness, leading to creep recovery. The maximum extent of the deformation by viscous flow depends on the glass volume fraction (GVF) and on the grain morphology [6].

LPS-ceramics can also deform through the mechanism of grain boundary sliding (GBS): rigid grains slide along the boundaries with their neighbours, which is facilitated by the lubricating amorphous layer. Small and equiaxial grains allow for a large GBS deformation. Elongated grains do not allow easy GBS deformation, since the rigid grains will rapidly interlock, giving rise to strain hardening [7]. Once grains interlock, a solution-diffusion-precipitation mechanism [8] will affect the creep rate. Due to this mechanism, neighbouring grains can continue to slide along their grain boundary [9], and secondary or steady-state creep is obtained. If the solution-diffusion-reprecipitation process is sufficiently rapid, GBS can lead to superplasticity [10].

Luecke et al. [11] and Lofaj et al. [12] have shown that, in the low-stress high-temperature creep regime, where ceramics are most likely to successfully compete with metallic materials, cavitation is a major contributor to tensile creep strain. In the model of Luecke and Wiederhorn [13] transport of matter to redistribute a deformable IGP is the critical step. The predicted creep rates are inversely proportional to the effective viscosity of the deformable phase and proportional to the cube of the volume fraction of the deformable phase.

Tertiary creep will occur if deformation is accelerated by damage processes such as cavitation coalescence and microcrack formation. Empirical relations between minimum creep rate and time to failure (Monkman-Grant), or between time to failure and stress (Larson-Miller) are established for individual materials. These relations suggest that creep resistance is improved by decreasing the amount and increasing the viscosity of the grain boundary phases [14,15].

Summarising the above, the volume fraction and viscosity of the amorphous IGP determine to a large extent an LPS ceramic's creep resistance. Unfortunately, these parameters cannot routinely be established from microstructural or microanalytical investigations. Methods to estimate these parameters in an indirect way have been proposed, some of them based on mechanical spectroscopy (MS).

MECHANICAL SPECTROSCOPY

Mechanical spectroscopy is the assessment of elastic and damping properties of materials using a cyclic stress or strain to probe a test sample. More in particular, mechanical spectroscopy investigates changes of elastic and damping properties as a function of temperature, and/or the frequency and amplitude of the sollicitation. Reference works on mechanical spectroscopy are by Nowick and Berry [16], and a new book edited by Schaller, Fantozzi and Gremaud [17].

Sudden changes of the elastic moduli with temperature are characteristic for particular microstructural events and are often accompanied by an increase of the damping. Damping is energy dissipation under cyclic loads, and is the manifestation of a stress-strain behaviour that differs from a purely elastic one. The damping property of a material is quantified in its "*specific damping capacity*", which equals the ratio $\Delta W/W$ of the energy dissipated in a full load cycle (ΔW) to the maximum stored energy during that cycle (W). Other terms to describe the same effect are "mechanical loss" and "internal friction". The phenomena encountered in this paper are most suitably described in terms of internal friction, and the corresponding symbol, Q^{-1}, will be used. One can measure Q^{-1} using one of the following relations.

$$Q^{-1} = \frac{1}{2\pi} \frac{\Delta W}{W_{el,\max}} = \tan \phi = \frac{k}{\pi f} \qquad \qquad Eq.1$$

For symmetric uniaxial, tension-compression tests, the energy dissipated in one cycle ΔW is directly obtained as the area of the stress-strain loop. $W_{el,\max}$, the maximum stored elastic strain energy in this cycle, equals $\frac{\sigma_{\max}^2}{2E}$. Q^{-1} can also be calculated from the phase angle ϕ between stress and strain, measured in e.g. a forced torsion pendulum. In addition, Q^{-1} is related to k, the exponential decay parameter of the vibration of a sample or system in free resonance at frequency f.

RESULTS OF MECHANICAL SPECTROSCOPY TESTS ON A LIQUID-PHASE-SINTERED SILICON CARBIDE

Figure 1 shows mechanical spectroscopy results obtained on a LPS-SiC material. These data will be used to illustrate different theories and interpretations reviewed in this paper. The results have been obtained using the impulse excitation technique (IET) [18], which is based on the analysis of the free-free bending vibration mode of a small rectangular test sample [nominally $6 \times 1 \times 30$ mm^3], and have been discussed in less detail earlier [19]. The test material was prepared by hot-pressing a mixture of SiC-powder, with 6wt% Al_2O_3 and 4wt% Y_2O_3 additives, as described by Sciti et al. [20]. After hot pressing, some of the material was subjected to an annealing treatment at 1900°C for 3 h in argon.

Figure 1 compares the LPS-SiC before and after the annealing heat treatment. For the as-hot-pressed SiC, a Q^{-1}-peak is observed, superimposed on a background, which increases monotonically with temperature. The annealing heat treatment at 1900°C has completely eliminated the Q^{-1}-peak, and the associated loss of stiffness. It is important to note that the IET-test, which involves heating up the sample at 2°C/min to 1400°C in N_2, leaves the shape and position of the Q^{-1}-peak of the as-hot-pressed material essentially unaffected. (Actually, the data shown in Figure 1 have been obtained during the first cooling stage.) This excludes the interpretation of the peak as a transitory irreversible effect associated with, e.g., crystallisation. Only phenomena that do not irreversibly alter the microstructure need to be considered when looking for its cause. First we will consider grain boundary sliding, in which case the peak position is related to the grain boundary viscosity.

Silicon-Based Structural Ceramics

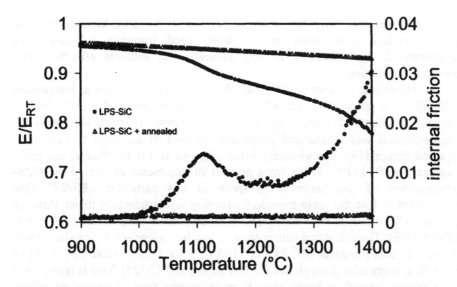

Figure 1 Comparison of high temperature Young's modulus and Q^{-1} for an as-hot-pressed and an annealed LPS-SiC.

ESTIMATION OF THE GRAIN BOUNDARY VISCOSITY FROM AN INTERNAL FRICTION PEAK DUE TO GRAIN BOUNDARY SLIDING

Raj and co-workers [21,22] were the first to investigate Q^{-1} in silicon nitride. They revealed a Q^{-1}-peak between 800 and 900 °C ($f = 0.73$ Hz). The peak was interpreted as the GBS-peak, by analogy to the GBS-peak in polycrystalline metals. The grain boundary sliding model is also supported by the work of Schaller and Lakki on oxide ceramics with and without amorphous intergranular films [23]. To investigate high-purity model silicon nitrides and carbides, where the GBS-peak is expected to occur at temperatures well above 1400 °C, Ota and Pezzotti developed a dedicated torsion pendulum. A small furnace can heat the part of the test sample between the grips to very high temperatures (over 2000 °C). Their paper in these proceedings [24] presents a summary of their findings and their most recent developments in test strategy.

Pezzotti proposed a procedure to estimate the grain boundary viscosity, η, at the GBS-peak temperature from the applied frequency [25]. This procedure has been applied successfully to a number of high-purity or intentionally doped silicon nitride and carbide materials. Thanks to an in-depth microanalysis of the grain boundary composition in the tested model materials, it was possible to compare the estimated grain boundary viscosities with the values of bulk glass viscosities reported in literature, providing excellent agreement. Since the knowledge of the grain boundary composition of commonly used materials is most often undetermined, this procedure is very interesting to estimate effective grain boundary viscosities.

To investigate whether the Q^{-1}-peak shown in Figure 1 for the as-hot-pressed LPS-SiC, can possibly be related to the GBS-mechanism, the procedure of Pezzotti to estimate the grain boundary viscosity was followed. Details of the microstructural and mechanical properties required to calculate η have been reported earlier [19]. The viscosity value obtained is 1.0×10^5 Pa.s at the peak-temperature of 1112°C. There are a number of arguments to discard the GBS-interpretation of this particular Q^{-1}-peak in this particular LPS-SiC. One observation is that the grain boundary viscosity is considerably lower than the viscosity of the IGP in BaO-doped SiC at a comparable temperature (10^9 Pa.s at 1100°C) [26]. The calculated values are also low in comparison to the viscosity of YSiAlON glass materials (10^8 Pa.s at 1100°C, 5.10^6 Pa.s at 1100°C) [27,28] or SiAlON intergranular glass phases (5.10^7 Pa.s at 1100°C) [29]. This is unexpected since silicon oxycarbide based glass is more viscous than an equivalent silicon oxynitride glass [26,30,31]. Most importantly, it seems unlikely that the GBS-peak occurs at a temperature where the grain boundary viscosity has decreased to 10^5 Pa.s. All of the viscosities at the Q^{-1}-peak temperature reported by Pezzotti et al. are in the range of 10^7-10^{10} Pa.s [26,24].

INTERNAL FRICTION PEAK ASSOCIATED WITH SOFTENING OF INTERGRANULAR GLASS POCKETS

Another phenomenon, which does not irreversibly alter the microstructure, and therefore could explain the observed damping peak (Figure 1) is the gradual softening of amorphous intergranular volumes.

Microstructural evidence for the 'softening-glass-pocket' interpretation

Sakaguchi et al. [32] were the first to report that in consecutive high temperature tests on a single specimen, the height of a Q^{-1}-peak observed near 1000°C decreases, and in some cases the peak disappears. Sakaguchi tentatively called upon the crystallisation of intergranular amorphous pockets to explain the disappearance of the peak. More systematic results were obtained by Lakki et al. [33]. TEM-investigations [34] showed that the disappearance of the Q^{-1}-peak reported by Lakki, is accompanied by the crystallisation of the amorphous pockets. In fact, after having increased the temperature to induce melting of the crystalline grain boundary phase, followed by rapid cooling to avoid crystallisation, the Q^{-1}-peak is restored. A similar link between the presence of amorphous pockets and the observation of a Q^{-1}-peak was provided by Roebben et al. for a sintered silicon nitride [35].

The comparison of the results in Figure 1 with the results of a high-resolution transmission electron microscopy (HRTEM) investigation (Figure 2, Figure 3) shows that the origin of the Q^{-1}-peak in the as-hot-pressed LPS-SiC indeed could be linked with the presence of amorphous pockets. Figure 2 proves the presence of amorphous and partially amorphous triple junctions in the as-sintered LPS-SiC. EDX-spectra reveal no evidence of other elements than those intentionally added in the material. The amorphous phase is rich in Y. The intergranular crystalline phase is rich in Al. After annealing, the multi-grain and triple junctions are fully crystalline, as shown in Figure 3.

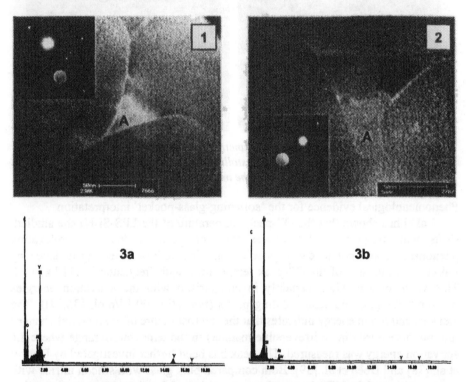

Figure 2 Dark-field images (ojective lens aperture positions are indicated) of two triple junctions (1 and 2) observed in the as-hot-pressed LPS-SiC, revealing (in 1) the amorphous phase (A) and (in 2) the coexistence of an amorphous (A) and a crystalline (C) phase in the intergranular pockets. Spectrum 3a shows the composition of the amorphous material (A) as observed in both pockets, whereas 3b shows that of the crystalline area (C) in (2)

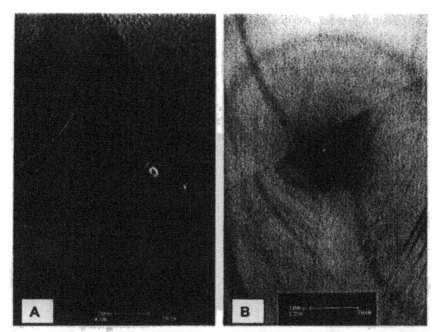

Figure 3 (HR)TEM image (A) of a triple junction (image B) observed in the hot-pressed and annealed LPS-SiC, indicating full crystallisation (evidenced by the lattice fringes in image A) of the material in the intergranular triple junctions.

Phenomenological evidence for the 'softening-glass-pocket' interpretation

Lakki has shown that the Q^{-1}-peak temperature of the LPS-Si$_3$N$_4$ she studied, shifts with frequency. This proves that the peak is due to a relaxation phenomenon, and not to e.g. crystallisation. The activation energy deduced by Lakki from the shift of the Q^{-1}-peak temperature with frequency is 1117 kJ/mol. This value is very high, certainly in comparison with the activation energies measured for grain boundary sliding peaks (typically 400 kJ/mol, [36,37]). The elevated activation energy indicates that the microstructure of the material changes (but not necessarily in an irreversible manner) in the temperature range where the activation energy was measured. The peak has been further investigated by Donzel et al. [38] and Doën et al. [39]. Both compared bulk glass with silicon nitride with a composition of the IGP close to the composition of the glass. Donzel reports that parameters describing the softening of a YSiAlON glass throughout its glass transition, i.e. the apparent activation energy ($E_{act} = 1116 \pm 20$ kJ/mol) and the so-called hierarchical correlation parameter ($b = 0.48 \pm 0.01$), are identical to those associated with the Q^{-1}-peak of the LPS-Si$_3$N$_4$ [38]. This suggests that the damping peak observed in this particular LPS-Si$_3$N$_4$, is related with the glass transition of its IGP. Indeed, the Q^{-1}-peak temperature is close to the glass transition temperature (range) of the intergranular glassy phase. Even if the peak temperature does not coincide with the glass transition temperature [31], the peak will be further referred to as the **T$_g$-peak**.

IET-tests revealed that the peak for the hot-pressed LPS-SiC shown in Figure 1 shifts with frequency. This required the use of several test samples with different size and consequently different resonant frequencies. The obtained value for the activation energy is 700 kJ/mol [40]. The frequency range however was limited to about 1 order of magnitude, and the correspondingly small shifts of the peak temperature affect the accuracy of the activation energy (R-value for linear fit = 0.74 [40]). While the shift of the peak with frequency proves the peak is due to a relaxation effect, the calculated activation energy fails to unequivocally associate the peak with the GBS- or T$_g$-peak interpretations.

ESTIMATION OF THE GLASS VOLUME FRACTION (GVF) FROM THE DECREASE OF YOUNG'S MODULUS ASSOCIATED WITH THE T$_G$-PEAK

The T$_g$-peak interpretation necessarily requires the height of the Q^{-1}-peak as well as the associated decrease of the E-modulus to correlate with the glass volume fraction (GVF). Roebben et al have shown that indeed the height of the Q^{-1}-peak of a particular LPS-Si$_3$N$_4$ is decreased after partial crystallisation of the IGP [35]. The relation between E-modulus and GVF is easily understood in terms of the composite nature of the LPS ceramic, with a rigid ceramic matrix phase and a second phase of glass channels and pockets. Recently, we presented [19,41] the following "GVF-procedure" to obtain an estimate of the GVF of a LPS-ceramic from the drop in stiffness at a T$_g$-peak.

At temperatures beneath the low-temperature foot of the T$_g$-peak, the LPS ceramic is considered as a 2-phase material, with a SiC or Si$_3$N$_4$ matrix phase and a solid, linear-elastic glassy phase. The stiffness of a material consisting of two solid phases is related to the stiffness of the individual phases and their volume fraction. Upper and lower boundary values for the composite stiffness can be determined from Voigt and Reuss approximations, as was done by Besson et al. investigating crystallisation of YSiAlON glass ceramics [42]. A more narrow yet accurate set of boundary values for materials containing a discontinuous second phase, is provided by Ravichandran [43]. In terms of the LPS-ceramic case, Ravichandran's analytical equations estimate the stiffness of the composite at the low-temperature foot of the Q^{-1}-peak (E$_{LT}$) as follows:

Lower bound:
$$E_{LT} = \frac{\left[(c\,E_{glass}\,E_{matrix} + E_{matrix}^2)(1+c)^2 - E_{matrix}^2 + E_{glass}\,E_{matrix}\right]}{(cE_{glass} + E_{matrix})(1+c)^2} \quad (3)$$

Upper bound:
$$E_{LT} = \frac{\left[E_{glass}\,E_{matrix} + E_{matrix}^2(1+c)^2 - E_{matrix}^2\right](1+c)}{(E_{glass} - E_{matrix})c + E_{matrix}(1+c)^3} \quad (4)$$

Upper and lower boundary values for E_{LT} can be calculated from the stiffness of the matrix material, E_{matrix}, the stiffness of the amorphous phase, E_{glass}, and c, a non-dimensional parameter related to the glass volume fraction, GVF:

$$c = \sqrt[3]{\frac{1}{GVF}} - 1 \quad (5)$$

At temperatures above the high-temperature foot of the Q^{-1}-peak, the glass phase has a very low stiffness and can be tentatively approximated as pores. In this case, the relation between the stiffness at the high-temperature foot of the Q^{-1}-peak (E_{HT}) and GVF is given by Eq. 6 [44]:

$$E_{HT} = E_{matrix}\,(1 - 1.9\,GVF + 0.9\,GVF^2) \quad (6)$$

One can combine equation (3) or (4) with (6) to eliminate E_{matrix}. However, between the lower and the higher temperature foot of the Q^{-1}-peak, E_{matrix} changes. Therefore, the values of E_{HT} and E_{LT} need to be corrected for the decay of matrix stiffness with temperature. The correction of E_{LT} is based on the slope of the E vs. T curve at temperatures below the Q^{-1}-peak, the correction of E_{HT} on the slope after this peak [19]. Substituting the measured and corrected values of E_{LT} and E_{HT}, one obtains an equation with GVF and E_{glass} as the only variables. A review of reported E-modulus values of Y-Si-Al-O-C glasses revealed a high consistency between individual results, allowing estimating the value of E_{glass} (at the temperature 'LT', below its glass transition) at 120 ± 10 GPa [42,45,46]. Hence, the value of GVF, the only unknown, can be calculated. The simple arithmetic average is selected for averaging the upper and lower bound values obtained by combining equation (6) with equations (3) and (4) respectively [44]. The 10 GPa estimation error on the value of E_{glass} corresponds with less than 1 vol% error on the calculated GVF.

It is often impossible to obtain independent confirmation of the calculated GVF. TEM does allow distinguishing amorphous and crystalline pockets. However, TEM lacks the statistical reliability and averaging capacity required for

GVF-estimation. SEM image analysis methods on the other hand lack the information about the crystallinity of the IGP. However, tests have been performed on a commercial LPS-silicon nitride (LS200 B from CeramTec, Plochingen, Germany), currently investigated in a Reference Material Test Programme of the ESIS TC6 Ceramics [47]. Contrary to the LPS-SiC material discussed in this paper, the LS200 B material does not contain crystalline intergranular phases, nor does it form them when tested up to 1400°C. The results of the GVF-procedure for LS200 B could therefore be compared with SEM image analysis, and excellent agreement was found, with values of 11.8 ± 2.1 vol% for the image analysis and 12.4 ± 0.6 vol% for the GVF procedure [41].

Details of the calculations for the LPS-SiC material, as well as for three other silicon based ceramics, are shown in [19]. An estimate of GVF of 5.8 vol% is obtained for the as-hot-pressed LPS-SiC. Compared with the SiO_2-content of the starting powders (1.4 wt%), the calculated GVF is rather large. However, undoubtedly part of the 6 vol% of Al_2O_3 and Y_2O_3 additives have entered the intergranular glass. Also, during processing the O-content might have increased. Therefore, the GVF-procedure seems to provide additional evidence for the interpretation of the peak shown in Figure 1 as being due to the softening of amorphous intergranular matter.

CURRENT ISSUES IN MECHANICAL SPECTROSCOPY OF LIQUID-PHASE-SINTERED CERAMICS

Industrial implementation and standardisation

The above sections have shown that high temperature mechanical spectroscopy has the potential to estimate otherwise inaccessible microstructural (glass volume fraction) and micromechanical (grain boundary viscosity) parameters. The procedures presented appear sound, and have been confirmed by direct comparison of their results with parallel microstructural or microanalytical investigations. It seems therefore interesting to implement the techniques in industrial practice. Given the small amount of test material and the relatively limited amount of time required to perform the tests, mechanical spectroscopy can be used in the early stages of the development of new high temperature resistant silicon-based ceramics. In addition, the use of non-destructive mechanical spectroscopy techniques, such as the Impulse Excitation Technique, allows testing final products for quality and process control [48].

However, few are the laboratories with particular mechanical spectroscopy facilities allowing reaching the very elevated temperature range where LPS silicon-based ceramics creep. Given a complete lack of standard test procedures, and small but often important differences between individual test apparatus, tests are, to the author's best knowledge, rarely duplicated or verified on an interlaboratory base. Bearing in mind the cautioning conclusions of the international round robin efforts co-ordinated by Luecke [49,50] on the subject of creep in silicon nitride, a similar action in the field of mechanical spectroscopy would be welcome. Only then, the technique can become a standard and recognised tool in the analysis of existing and newly developed LPS-ceramics.

On the other hand, there are a number of issues in the field of structural ceramics for high temperature applications that require more in-depth and fundamental mechanical spectroscopy research. The following paragraphs speculate on some critical points in this matter.

The co-occurrence of GBS- and T_g-peaks

So far, no LPS-ceramic has been reported to display both a T_g-peak and a GBS-peak. The reasons for this observation are not yet identified. A most tempting interpretation would be that the presence of one of the peaks excludes the occurrence of the other. It might be speculated that the presence of a significant amount of amorphous and soft volumes, inducing the T_g-peak, reduces the elastic restoring force required to induce the GBS-peak. For such materials, the GBS peak could be difficult to distinguish from the viscoplastic background.

The "exponential" background damping

The contribution to creep strain stemming from anelastic deformation mechanisms, which induce internal friction peaks, will be recovered after unloading. Fundamentally different is the viscoplastic part of creep deformation, with a permanent, plastic character. The continuous accumulation of plastic deformation requires an accommodation process, such as solution-diffusion-reprecipitation, or cavitation. Depending on the material, temperature, stress amplitude, but also the test type (creep or internal friction), the irreversible accommodating mechanism or the reversible GBS-mechanism itself are strain rate-limiting.

The plastic, unrecoverable part of creep strain will not induce a Q^{-1}-peak. Nevertheless, in the process of plastic deformation, applied strain energy will be dissipated, and under cyclic loading conditions this leads to internal friction. The viscoplastic strain and Q^{-1} will increase with temperature, and result in an exponentially increasing background damping. The contribution of creep according to a given creep law to Q^{-1} can be easily calculated. Both Pezzotti et al. [25] and Lakki et al. [51] have provided evidence that the background Q^{-1} at elevated temperatures can be predicted from low stress creep data, performed in situ in the torsion pendulum.

Silicon-Based Structural Ceramics

The stress amplitude issue

Unfortunately, there is a gap between the stress levels reached in torsion pendulums (shear stress limited to 10 or 15 MPa), or resonant vibration techniques ($\sigma_a \ll 1$ MPa), and the stresses realistically encountered in industrial applications. Also, the generally observed tension-compression creep asymmetry is not taken into account or revealed by the reported torsion pendulum test results. Higher stress amplitudes can be reached in the uniaxial mode. Equipment, which allows reversed loading, can be used to measure high amplitude internal friction, as well as the tension-compression asymmetry. Roebben et al. reported a difference between low-amplitude torsion pendulum and high-amplitude uniaxial tension-compression test results obtained on a LPS-Si_3N_4 containing Y- and Al-additives [35,52]. Q^{-1} is stress-amplitude independent at low σ_a. Above a temperature and frequency dependent threshold stress amplitude (50 MPa at 1300°C and 0.01 Hz), a second damping component is superimposed on the low-amplitude component, leading to a much larger and stress-amplitude dependent energy dissipation.

The hypothesis of non-Newtonian viscous behaviour was investigated. Often, the stress-amplitude dependence of creep rates is taken into account by adopting a Norton-type stress-strain rate relation with stress exponent n ($\dot{\varepsilon} \sim \sigma^n$). When a material obeying the Norton law is periodically loaded, a power law relation with exponent n-1 is established between Q^{-1} and σ_a [51]. This finding was incorporated in a rheological spring-dashpot model [35]. According to this very general model the total deformation is the sum of a linear-elastic component (spring S1), a fully recoverable anelastic component (parallel combination of spring S2 and dashpot D1), and a viscous irrecoverable deformation (dashpot D2). The stress-strain rate relationship of the dashpots was imposed to be $\dot{\varepsilon} = K.\sigma^n$, with an exponent n of 2, as this is required to obtain the linear relationship between ΔW and σ_a observed at large σ_a.

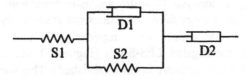

Figure 4 Spring-dashpot model describing the high-T deformation of a LPS-ceramic

The steady-state σ-ε-loop, with the shape of a closed ellipse, is satisfactorily simulated by a Maxwell-type two-parameter model consisting of a serial combination of an elastic spring (S1) and a dashpot (D2) [35]. However, the first applied cycle results in an open loop (Figure 5). This shape cannot be simulated with a serial spring-dashpot model. The loop can be described with a Voigt-type two-parameter model, i.e. a parallel combination of a spring (S2) and a dashpot (D1). For the total deformation the three-element-model (S1, S2 and D1), which is the standard anelastic model, was assumed. The stiffness of S1 was determined from cyclic tests at a higher frequency (1 Hz). To S2 a stiffness E_{an} equal to the stiffness E of spring S1 was attributed. The dashpot-constant K was varied iteratively until the width of the calculated loop agrees with the experimentally determined loop, resulting in a satisfactory simulation (Figure 5).

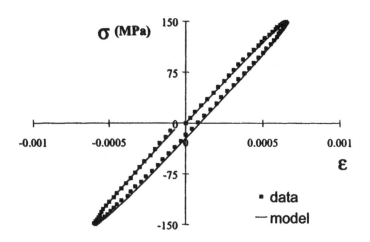

Figure 5 σ-ε-loop (1st of a series) obtained on a LPS-Si$_3$N$_4$ at 1300°C, at a frequency of 0.01 Hz; solid line is calculated from a series combination of a spring (E = 245 GPa) and a parallel spring-dashpot system (K = 2 . 10^{-10} (MPa)$^{-2}$s^{-1}, E$_{an}$ = 245 GPa).

As this rheological analysis indicates anelastic behaviour, one can conclude that the high-amplitude energy dissipation does not stem solely from viscoplastic creep phenomena. It has indeed been reported [53] that part of the creep deformation of the investigated LPS-Si$_4$N$_4$ (Figure 5) can be recovered upon unloading (although a full recovery is not established). This was also observed by Croizet [7], who measured and modelled the creep behaviour at 1250°C of this particular LPS-Si$_3$N$_4$. The stress exponent of Croizet's phenomenological model was 3.7. Gasdaska [54] and later Li and Reidinger [55] investigated another LPS-Si$_3$N$_4$, with a similar amount of IGP. A stress exponent for secondary, steady-state creep of 3.2 ± 0.2 was found. However, the stress exponent value of the initial, recoverable creep, which occurs at a higher rate was 1.8 ± 0.4, similar to the stress exponent used in the simulation of Figure 5. This suggests that the mechanism causing the stress-amplitude contribution to Q^{-1}, as shown in Figure 5 is related to the initial and recoverable part of the creep deformation.

The origin of the large recoverable strain is still under investigation. Gasdaska [54] recognised the importance of the difference in elastic properties between silicon nitride and crystalline secondary phases. Due to its lower stiffness the IGP experiences a lower stress upon loading. As GBS allows the grains to rearrange the stresses are redistributed. The latter process is reversible and would indeed lead to an anelastic effect. Li and Reidinger [55] have suggested a similar stress redistribution, from the faster creeping small-grained matrix to the larger grains in a bimodal, self-reinforced silicon nitride.

Elevated temperature fatigue of liquid-phase-sintered ceramics

Several authors have reported that at high temperatures some ceramics are more susceptible to failure under static load than under cyclic load. Amongst the earlier observations are those of Bolsch [56], who tested a hot-pressed silicon nitride under uniaxial tension-compression load at 1200 °C. His data suggest the existence of a fatigue threshold stress-amplitude of 320 MPa for the investigated load-ratio R = -1. Similar positive fatigue effects have been observed more often in crack propagation tests.

A review of the fatigue properties of non-transforming monolithic ceramics was published in 1996 [57]. At low temperature the fatigue effect is cycle dependent, i.e. related to the number of crack opening and crack closure events. Frequency dependent phenomena determine the fatigue effect at high temperature. Specifically, the fatigue effect at high temperature depends on the amount and viscosity of the amorphous fraction of the IGP. When an amorphous IGP is present, the mechanical behaviour and the damage susceptibility of the ceramic become very strain rate sensitive. Room temperature fatigue mechanisms which cause the crack growth rate to increase relative to that under static loading, tend to be suppressed at high temperatures because of the overriding potential of the IGP to accommodate morphological mismatches due to its low viscosity.

The results shown in Figure 5 explain at least partially the observations of a positive fatigue effect. The answer is found in the truly anelastic deformation contribution, which dissipates large quantities of energy, particularly at large stress amplitude. The dissipation of large amounts of energy in an anelastic hence non-damaging way, inevitably renders a material more resistant to fast cycling loads.

High temperature toughness of liquid-phase-sintered ceramics

Donzel [28] has shown that the occurrence of a T_g-peak in zircon ceramics ($ZrSiO_4$) is reflected in a fracture toughness peak. The LPS-Si_3N_4 material investigated here (Figure 5) displays a similar behaviour, with a small Q^{-1}-peak at 980°C and 1 Hz [35], and a K_c-peak at about 1100°C (Figure 6, [7]).

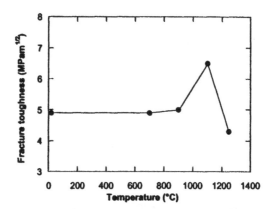

Figure 6 Fracture toughness of LPS-Si₃N₄ as a function of temperature (from bending tests on Vickers indented samples by Croizet [7])

Although the K_c-data lack a good temperature resolution, they suggest that the K_c-peak is located at higher temperatures than the T_g-peak. This agrees with the time-temperature equivalence, as the fracture toughness was measured at higher loading rates (0.8 mm/min = 90 MPa/s) than that applied during the torsion pendulum test (f = 1 Hz, σ_a = 4 MPa, hence, average loading rate is 16 MPa/s).

The stress amplitude dependence of the high amplitude energy dissipation, which was observed to be of an anelastic nature, is particularly interesting for the understanding of the K_c-peak (Figure 6). The stress amplitude in a cracked component submitted to cyclic loads will be maximum at the crack tip. The linear relationship between Q^{-1} and stress amplitude implies that the energy dissipation will automatically be largest at the place where energy dissipation is most advantageous, i.e. at the crack tip.

It is of utmost importance to recognise that the positive effect on toughness can only occur simultaneously with a satisfactory creep resistance, if the underlying energy dissipation is of anelastic and not of plastic nature. The particularly large Q^{-1}-contribution at high stress amplitudes was suggested to be due to the recoverability of the delayed deformation of the percolated network-structure of silicon nitride grains [58]. Since this structure has been recognised to be very creep resistant [59], it is currently the most promising kind of microstructure for high quality silicon-based ceramics.

CONCLUSIONS

This paper demonstrated how mechanical spectroscopy techniques can be used to assess microstructural and micromechanical parameters critical for the high temperature behaviour of liquid phase sintered ceramics. Particular Q^{-1}-peaks have been associated with grain boundary sliding and glass softening phenomena. Procedures to assess the intergranular glass volume fraction and the grain boundary viscosity have been outlined. These parameters can be used in the modelling and assessment of the high temperature behaviour of liquid-phase-sintered ceramics. More direct relations between the results of mechanical spectroscopy tests and properties such as creep and fatigue resistance and fracture toughness are less established, while highly desired.

ACKNOWLEDGEMENTS

The authors thank Diletta Sciti and Alida Bellosi (IRTEC-CNR, Faenza, Italy) for having provided the LPS-SiC samples, and Marc Steen (JRC-IE, Petten, the Netherlands) for the guidance in the uniaxial tension-compression tests. Gert Roebben is a post-doctoral fellow of the FWO-Vlaanderen.

REFERENCES

1. D. R. Clarke, "On the Equilibrium thickness of intergranular glass phases in ceramic materials", *J. Amer. Ceram. Soc.*, **70** 15-22 (1987).
2. H.-J. Kleebe, "Structure and chemistry of interfaces in Si_3N_4 ceramics studied by transmission electron microscopy," *J. Ceram. Soc. Jpn.*, **105** 453-475 (1997).
3. J.R. Dryden, D. Kucerovsky, D.S. Wilkinson, D.F. Watt, "Creep deformation due to a viscous grain boundary phase," *Acta metall.*, **37** 2007-2015 (1989).
4. M.M. Chadwick, R.S. Jupp, D.S. Wilkinson, "Creep behavior of a sintered silicon nitride," *J. Amer. Ceram. Soc.*, **76** [2] 385-396 (1993).
5. Q. Jin, X.-G. Ning, D. S. Wilkinson, G. Weatherly, "Redistribution of a grain-boundary glass phase during creep of silicon nitride ceramics," *J. Amer. Ceram. Soc.* **80** 685-691 (1997).
6. J. R. Dryden, D. S. Wilkinson, "Three-dimensional analysis of the creep due to a viscous grain boundary phase," *Acta mater.* **45** 1259-1273 (1997).
7. D. Croizet, "Etude experimentale et numerique du comportement a haute temperature d'un nitrure de silicium," Dissertation, Ecole Nationale Supérieure des Mines de Paris, 1992.

8. R. Raj, C. K. Chyung, "Solution-precipitation creep in glass ceramics," *Acta metall.* **29** 159-166 (1981).
9. R. L. Tsai, R. Raj, "Overview 18: Creep fracture in ceramics containing small amounts of a liquid phase," *Acta metall.* **30** 1043-1058 (1982).
10. T. Rouxel, F. Rossignol, J.-L. Besson, P. Goursat, "Superplastic forming of an alpha-phase rich silicon nitride," *J. Mater. Res.* **12** 480-492 (1997).
11. W. E. Luecke, S. M. Wiederhorn, B. J. Hockey, R. F. Krause Jr., G. G. Long, "Cavitation contributes substantially to tensile creep in silicon nitride," *J. Amer. Ceram. Soc.* **78** 2085-2096 (1995).
12. F. Lofaj, A. Okada, H. Kawamoto, "Cavitational strain contribution to tensile creep in vitreous bonded ceramics," *J. Amer. Ceram. Soc.* **80** 1619-1623 (1997).
13. W. E. Luecke, S. M. Wiederhorn, "A new model for tensile creep of silicon nitride," *J. Amer. Ceram. Soc.* **82** 2769-2778 (1999).
14. S. M. Wiederhorn, B. J. Hockey, D. C. Cranmer, R. Yeckley, "Transient creep behaviour of hot isostatically pressed silicon nitride," *J. Mater. Sci.* **28** 445-453 (1993).
15. J.-L. Ding, K.C. Liu, K.L. More, C.R. Brinkman, "Creep and creep rupture of an advanced silicon nitride ceramic," *J. Amer. Ceram. Soc.* **77** 867-874 (1994).
16. A. S. Nowick, B. S. Berry, "Anelastic relaxation in crystalline solids," Academic Press, New York (1972)
17. "Mechanical spectroscopy, Q^{-1} 2001, with applications to materials science," ISBN 0-87849-876-1, ed. R. Schaller, G. Fantozzi and G. Gremaud, Trans Tech Publications Ltd, Switzerland (2001).
18. G. Roebben, B. Bollen, A. Brebels, J. Van Humbeeck, O. Van der Biest, "Impulse excitation apparatus to measure resonant frequencies, elastic moduli and internal friction at room and high temperature," *Rev. Sci. Instrum.* **68** 4511-4515 (1997).
19. G. Roebben, R.G. Duan, D. Sciti, O. Van der Biest, "Assessment of the high temperature elastic and damping properties of silicon nitrides and carbides with the impulse excitation technique (IET)," *J. Europ. Ceram. Soc.* **22** [14-15] 65-73 (2002).
20. D. Sciti, S. Giucciardi, A. Bellosi, "Effect of annealing treatments on microstructure and mechanical properties of liquid-phase sintered silicon carbide," *J. Eur. Ceram. Soc.* **21** 621-632 (2001).
21. D. R. Mosher, R. Raj, R. Kossowsky, "Measurement of viscosity of the grain-boundary phase in hot-pressed silicon nitride," J. Mater. Sci. **11** 49-53 (1976).
22. R. L. Tsai, R. Raj, "The role of grain-boundary sliding in fracture of hot-pressed Si_3N_4 at high temperatures," *J. Amer. Ceram. Soc.* **63** 513-517 (1980).

23. R. Schaller, A. Lakki, "Grain boundary relaxations in ceramics"; pp. 315-337 in *Mechanical spectroscopy, Q^{-1} 2001, with applications to materials science*, ISBN 0-87849-876-1, ed. R. Schaller, G. Fantozzi and G. Gremaud, Trans Tech Publications Ltd, Switzerland, 2001.

24. G. Pezzotti, K. Ota, "Grain-boundary relaxation processes in silicon-based ceramics studied by mechanical spectroscopy," these proceedings.

25. G. Pezzotti, K. Ota, H.-J. Kleebe, Y. Okamoto, T. Nishida, "Viscous behavior of interfaces in fluorine-doped Si$_3$N$_4$/SiC composites," *Acta metall. mater.* **43** 4357-4370 (1995).

26. G. Pezzotti, H. Nishimura, K. Ota, H.-J. Kleebe, "Grain-boundary viscosity of BaO-doped SiC," *J. Amer. Ceram. Soc.* **83** 563-570 (2000).

27. T. Rouxel, P. Verdier, "SiC particle reinforced oxynitride glass and glass-ceramic composites: crystallization and viscoplastic forming ranges," *Acta mater.* **44** 2217-2225 (1996).

28. L. Donzel, "Intra- and intergranular high temperature mechanical loss in zirconia and silicon nitride," Dissertation, Ecole Polytechnique Fédérale de Lausanne, 1998.

29. G. Pezzotti, H.-J. Kleebe, K. Okamoto, K. Ota, "Structure and viscosity of grain boundary in high-purity SiAlON ceramics," *J. Amer. Ceram. Soc.* **83** 2549-2555 (2000).

30. G. Pezzotti, H.-J. Kleebe, K. Ota, "Grain-boundary viscosity of polycrystalline silicon carbides," *J. Amer. Ceram. Soc.* **81** 3293-3299 (1998).

31. T. Rouxel, J.-C. Sangleboeuf, M. Huger, C. Gault, J.-L. Besson, S. Testu, "Temperature dependence of Young's modulus in Si$_3$N$_4$-based ceramics: roles of sintering additives and of SiC-particle content," *Acta Mat.* **50** 1669-1682 (2002).

32. S. Sakaguchi, N. Murayama, F. Wakai, "Internal friction of Si$_3$N$_4$ at elevated temperatures," *J. Ceram. Soc. Jpn. Inter. Ed.* **95** 1162-1165 (1987).

33. A. Lakki, R. Schaller, "Anelastic relaxation associated with the intergranular phase in silicon nitride and zirconia ceramics," *Journal of Alloys and Compounds* **211/212** 365-368 (1994).

34. A. Lakki, R. Schaller, G. Bernard-Granger, R. Duclos, "High temperature anelastic behaviour of silicon nitride studied by mechanical spectroscopy," *Acta metall. mater.* **43** 419-426 (1995).

35. G. Roebben, L. Donzel, S. Stemmer, M. Steen, R. Schaller, O. Van der Biest, "Viscous energy dissipation in silicon nitride at high temperatures," *Acta mater.* **46** 4711-4723 (1998).

36. G. Pezzotti, H.-J. Kleebe, K. Ota, "Grain-boundary viscosity of polycrystalline silicon carbides," *J. Amer. Ceram. Soc.* **81** 3293-3299 (1998).

37. G. Pezzotti, "Grain-boundary viscosity of Calcium-doped silicon nitride," *J. Amer. Ceram. Soc.* **81** 2164-2168 (1998).

38. L. Donzel, A. Lakki, R. Schaller, "Glass transition and α-relaxation in Y-Si-Al-O-N glasses and in Si_3N_4 ceramics studied by mechanical spectroscopy," *Phil. Mag. A* **76** 933-944 (1997).

39. B. Doën, P. Gadaud, A. Rivière, "Silicon nitrides studied by isothermal mechanical spectroscopy," *J. Alloys Compd.* **310** 36-38 (2000).

40. G. Roebben, D. Sciti, C. Sarbu, T. Lauwagie, A. Bellosi, O. Van der Biest, "High temperature stiffness and damping measurements to monitor the glassy intergranular phase in liquid-phase-sintered silicon carbide", submitted to J. Am. Ceram. Soc. (2002).

41. G. Roebben, C. Sarbu, T. Lube, O. Van der Biest, "Quantitative determination of the volume fraction of intergranular amorphous phase in sintered silicon nitride," presented at the 13th ICIFUAS, Bilbao, July 2002, submitted to Mat. Sc. Engng A (2002).

42. J.-L. Besson, H. Lemercier, T. Rouxel, G. Trolliard, "Yttrium sialon glasses: nucleation and crystallization of Y35Si45Al20O83N17," *J. Non-Crystalline Solids* **211** 1-21 (1997).

43. K. S. Ravichandran, "Elastic properties of two-phase composites," *J. Amer. Ceram. Soc.* **77** 1178-1184 (1994).

44. D. J. Green, "An introduction to the mechanical properties of ceramics," Cambridge Solid State Science Series, Cambridge University Press, 1998.

45. R. Ramesh, P. Chevaux, H. Lemercier, M. J. Pomeroy, S. Hampshire, "Characterization of oxycarbide glasses prepared by melt solidification," Euro Ceramics V, Proceedings of the 5th Meeting of the European Ceramic Society, Versailles, *Key Engineering Materials* **132-136** 189-192 (1997).

46. T. Rouxel, J.-C. Sangleboeuf, P. Verdier, Y. Laurent, "Elasticity, stress relaxation and creep in SiC particle reinforced oxynitride glass," *Key Engineering Materials* **171-174** 733-740 (2000).

47. T. Lube, R. Danzer, M. Steen, "A testing program for a silicon nitride reference material," submitted to: Improved Ceramics Through New Measurements, Processing and Standards, Ceramic Transactions 133, Proc. of PacRim 4, No. 4-8 2001, Wailea, Hawaii, USA, 2002.

48. G. Roebben, B. Basu, J. Vleugels, J. Van Humbeeck, O. Van der Biest, "The innovative impulse excitation technique for high-temperature mechanical spectroscopy," *J. Alloys Compd.* **310** 284-287 (2000).

49. W. E. Luecke, S. M. Wiederhorn, "Interlaboratory verification of silicon nitride tensile creep properties," *J. Amer. Ceram. Soc.* **80** 831-838 (1997).

50. W. E. Luecke, "Results of an international round-robin for tensile creep rupture of silicon nitride," *J. Amer. Ceram. Soc.* **85** 408-414 (2002).

51. A. Lakki, R. Schaller, M. Nauer, C. Carry, "High temperature superplastic creep and internal friction of yttria doped zirconia polycrystals," *Acta metall. mater.* **41** 2845-2853 (1993).
52. G. Roebben, L. Donzel, M. Steen, R. Schaller, O. Van der Biest, "Fatigue resistant silicon nitride ceramics due to anelastic deformation and energy dissipation," *J. Alloys Compd.* **310** 39-43 (2000).
53. M. Bartsch, "Ribausbreitungsverhalten von Siliziumnitrid-Werkstoffen unter mechanischer Beanspruchung bei Raum- und Hochtemperatur," Dissertation, Technische Hochschule Darmstadt, 1996.
54. C.J. Gasdaska, "Tensile creep in an in situ reinforced silicon nitride," *J. Amer. Ceram. Soc.* **77** [9] 2408-2418 (1994).
55. C.-W. Li, F. Reidinger, "Microstructure and tensile creep mechanisms of an in situ reinforced silicon nitride," *Acta mater.* **45** 407-421 (1997).
56. D. Bolsch, "Untersuchung zum wechselfestigkeitsverhalten von siliciumnitrid bei hohen temperaturen," Dissertation, Karlsruhe Universität, 1990.
57. G. Roebben, M. Steen, J. Bressers, O. Van der Biest, "Mechanical fatigue in monolithic non-transforming ceramics," *Prog. Mater. Sci.* **40** 265-331 (1996).
58. G. Roebben, "Viscous energy dissipation in silicon nitride at high temperature," Dissertation, K.U.Leuven, 1999.
59. D. S. Wilkinson, "Creep mechanisms in multiphase ceramic materials," *J. Amer. Ceram. Soc.* **81** 275-299 (1998).

HIGH-TEMPERATURE DEFORMATION OF SILICON NITRIDE AND ITS COMPOSITES

G. A. Swift, E. Üstündag*, B. Clausen
California Institute of Technology
Materials Science, 138-78
Pasadena, CA 91125

M. A. M. Bourke
Materials Science and Tech. Division
Los Alamos National Laboratory
Los Alamos, NM 87545

H. T. Lin
Metals and Ceramics Division
Oak Ridge National Laboratory
Oak Ridge, TN 37831

C. W. Li
Corporate Research and Technology
Honeywell Corporation
Morristown, NJ 07962

ABSTRACT
Neutron diffraction was utilized in the investigation of elastic lattice strain evolution during high-temperature deformation of monolithic *in-situ* reinforced Si_3N_4 and its SiC-particle composite. Tension experiments were performed near 1400°C using the new SMARTS diffractometer at the Los Alamos Neutron Science Center. The diffraction data provided information about thermal expansion coefficients and elastic constants at high temperature. Also, the *hkl*-dependent strains were measured.

INTRODUCTION
        Structural ceramics have been gaining more use worldwide as their properties are enhanced and verified through the efforts of materials scientists. For materials to be used at elevated temperatures it is important, particularly for structural materials, to understand their mechanical properties at such temperatures. Silicon nitride, Si_3N_4, is among the most widely used ceramic materials. A very promising variety is AS800 (Honeywell Ceramic Components, Torrance, CA), which has an acicular grain structure providing *in-situ* reinforcement (ISR). Such a material is envisioned as being applicable for hostile environments such as turbine blades. ISR Si_3N_4 generally has good creep resistance, good strength retention at high temperature, and high fracture toughness [1]. This study was motivated by its commercial availability and by

---

* Corresponding author; E-mail: ersan@caltech.edu

---

the increased interest in such materials due to their technological importance. Besides an ISR-Si$_3$N$_4$, a Si$_3$N$_4$ sample containing silicon carbide particles (SiC) was tested. Such an addition to Si$_3$N$_4$ is believed to have beneficial effects on high-temperature properties, especially creep resistance, though the actual effect seems somewhat ambiguous [2-4].

Previous high-temperature tensile studies have been performed on Si$_3$N$_4$, but these have focused mainly on "traditional" creep studies [5,6]. Neutron diffraction can measure the bulk volume average response to applied strain in a sample while the strain is applied. In the present research, neutron diffraction was employed to measure such microstructural response to applied stress at high temperature. The present research is a verification of capability for a new instrument (SMARTS), and only elastic tensile load was applied.

High-temperature neutron diffraction was used in this research. There is only a small number of such studies in the literature, due to lack of suitable instrumentation. Among those few are relatively low temperature (~400°C) composite creep studies using neutron diffraction [7,8]. Typically in these studies, the diffraction strains were noted as nearly constant during the creep testing, as diffraction is capable of recording *elastic* strains only, while effects such as peak broadening are noted as possibilities during plastic deformation.

This report details the characterization of the high-temperature properties of AS800 and SiC$_p$-Si$_3$N$_4$, including the single crystal elastic constants and thermal expansion coefficient. This is the first in-depth investigation using neutron diffraction to quantify the high-temperature behavior of an ISR Si$_3$N$_4$ and its composite. The calculation of these parameters and comparison to literature demonstrates the capabilities of the SMARTS system.

EXPERIMENTAL PROCEDURE
Neutron Diffraction
The present study makes use of the SMARTS diffractometer at the Los Alamos Neutron Science Center [9]. SMARTS employs time-of-flight neutron diffraction, so a sample's entire diffraction pattern be obtained quickly, rather than using a constant-wavelength (reactor) source, which would limit data acquisition to one or only a few peaks. The neutron beam encompassed the entire gage width (centered at the midpoint of the gage length) of the sample. The sample was oriented in the loading fixture so that it made a 45° angle to the incident beam. Diffraction data were collected at Bragg angles of $2\theta = \pm 90°$. This corresponds to transverse (+90°) and longitudinal (-90°) diffraction strains.

Diffraction data were analyzed using the Rietveld method [10] via the GSAS program [11]. The Rietveld method is a least squares method that takes into account certain parameters affecting the diffraction pattern in order to fit the experimental data. Input parameters for full-pattern fits included the space group (p63/m), and literature values for lattice parameters, $a = 7.608$Å, $c = 2.911$Å for β-Si$_3$N$_4$, and $a = 4.361$Å for SiC (space group F43m) [12]. Refined parameters include the lattice parameters, thermal parameters, and preferred orientation. Average fitting errors were ~6%. Single peaks were fit with a Voigt function, with typical error ~8%. These single peak fits were used for *hkl*-dependent calculations.

Patterns were obtained for the samples at room temperature, the test temperature, and several intermediate temperatures. The patterns from these intermediate temperatures allowed for computation of the coefficient of thermal expansion. The AS800 patterns exhibit β-Si$_3$N$_4$ only, while the persistent phase resulting from the liquid-phase sintering manufacturing process does not appear, despite a pre-test heat treatment known to crystallize the majority of this persistent phase [6]. As a constituent component will appear if its amount is about 5% or more, this indicates the second phase is a small component indeed. The SiC$_p$-Si$_3$N$_4$ diffraction patterns were also primarily β-Si$_3$N$_4$; there was only one independent SiC peak, however, the rest overlapping with β-Si$_3$N$_4$.

Diffraction patterns were recorded every 15 minutes for the larger AS800 sample, every 30 minutes for the composite sample; these were the minimum times necessary for high quality diffraction patterns, in accord with similar methods [7]. The heating rate was 20°C/min, and each intermediate temperature was maintained until a high temperature extensometer indicated no further strain increase before diffraction patterns obtained, indicating temperature equilibration. Diffraction patterns for increasing stress levels were recorded in quick succession to note any changes during the test. Patterns obtained after unloading, while cooling, and after fully cooled were used to check for residual strain resulting from the testing.

Thermo-Mechanical Testing

Tensile tests were carried out in vacuum at elevated temperature using a horizontally aligned stress fixture. Tensile tests were performed at constant loads at elevated temperature on pin-loaded, dog-bone samples with a gage length of 51 mm. The AS800 sample had a gage section of 6.3 x 5.0 mm^2, while the composite sample was 3 x 3 mm^2. The loading fixture (Instron Corp., Canton, MA) and furnace (MRF Inc., Suncook, NH MRF) were custom-made for use in concert with neutron diffraction. The high temperature grips were W-10%Ta, while Al windows in the furnace allowed easy penetration of the neutron beam. Samples were gripped in the furnace hot-zone center, also the center of diffraction. A constant stress of 30 MPa was maintained during the heat-up and cool-down. The AS800 sample was heated to the test temperature of 1375°C, while the composite sample was tested at 1400°C. During the heating, a high-temperature extensometer measured sample extension.

Tensile load was increased after reaching the test temperature. The sample was subjected to successively increasing stresses to a maximum of 175MPa. Each load was held constant for 45-60 minutes while diffraction patterns were recorded. After about seven hours at high load and high temperature, the load was decreased to 125MPa and 75MPa for one scan each, and finally unloaded to the initial stress of 30MPa, and patterns were obtained for over 2 hours. The furnace was then cooled to room temperature and a final scan was performed. Only diffraction strains are presented, as an extensometer data acquisition error occurred in the early stages of increasing stresses

RESULTS AND DISCUSSION
Thermal Expansion

AS800: Since diffraction patterns were recorded at several temperatures during the heating process, it was possible to determine the thermal expansion

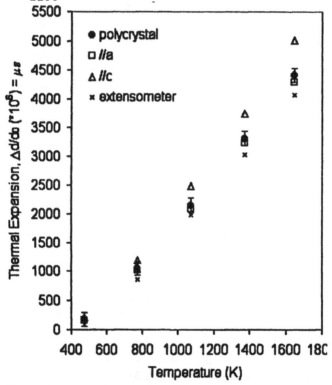

Figure 1. Thermal expansion of AS800 based on longitudinally measured lattice parameters. The CTE was calculated as 3.64 x 10^{-6} K^{-1}, the slope of a linear fit to the polycrystalline data.

coefficient of the sample. The full pattern fits from GSAS, and thus the average changes in the a and c lattice parameters, were used. The data shown in Figure 1 are longitudinal ($2\theta = -90°$) strains. The data points are the averages of the multiple scans from each temperature, with the strains relative to the initial room temperature scan. Figure 1 shows the change in each lattice parameter, and the polycrystalline average: $(2a + c)/3$ to be in close agreement with the extensometer data, which are also averaged over the time at each temperature. The thermal expansion is linear over the entire temperature range for the 30MPa applied nominal stress. A linear fit to the polycrystalline data gives $\alpha = 3.64 \times 10^{-6}$ K^{-1}, while the transverse data (not shown) yielded $\alpha = 3.57 \times 10^{-6}$ K^{-1}. These results agree well with the value quoted by Honeywell of $\alpha = 3.9 \times 10^{-6}$ K^{-1} up to 1000°C.

Si_3N_4 is hexagonal, so the α tensor components are parallel to the crystallographic axes. Thus can the CTE can be found as above, by changes in the lattice parameters [13]. Calculating these lattice parameters first, *then* finding the tensor components incurs an error [13]. Using the program ALPHA [13], the individual reflection positions were input and a least squares evaluation fit the measured data. Nine different reflections were fit as single peaks, giving d-spacings for each reflection as a function of temperature. All d-spacing data were used to determine the CTE tensor of AS800 over the full temperature range: $\alpha_{11} = \alpha_{22} = 3.50$ (±0.03) $\times 10^{-6}$ K^{-1}, $\alpha_{33} = 4.06$ (±0.04) $\times 10^{-6}$ K^{-1}. There are only two independent terms, since Si_3N_4 is hexagonal [14]. Using $\alpha = (2\alpha_{11} + \alpha_{33})/3$, gives 3.69×10^{-6} K^{-1}, similar to the above polycrystal result, obtained using specific peak positions rather than average lattice parameters computed from the full patterns. Since less error is propagated when calculating CTE directly from reflection positions [13], this CTE value is more accurate. Note that this CTE is for a processed AS800 sample, which properties are different from, say, a powder sample, especially because the acicular structure of ISR-Si_3N_4 only forms with processing. Using these tensor components, the degree of anisotropy was determined by calculating the aspherism index [14], A = 0.034, while perfect isotropy would be A = 0.

SiC_p-Si_3N_4: Figure 2 shows the thermal expansion of both phases in the composite sample. Linear fits give $\alpha = 3.35 \times 10^{-6}$ K^{-1} for the polycrystalline average of Si_3N_4 and $\alpha = 4.33 \times 10^{-6}$ K^{-1} for SiC. These values are comparable with literature values of 3.60×10^{-6} K^{-1} and 4.70×10^{-6} K^{-1} [15]. The likely reason for discrepancy is the peak overlap between the two phases. Single peak fits for Si_3N_4 reflections which were *not* overlapped with SiC reflections were used in ALPHA, giving $\alpha_{11} = \alpha_{22} = 3.38$ (±0.07) $\times 10^{-6}$ K^{-1} and $\alpha_{33} = 4.21$ (±0.10) $\times 10^{-6}$ K^{-1}. Computing the polycrystalline average of these two (using α_{11} for a and α_{33} for c) gives $\alpha = 3.66 \times 10^{-6}$ K^{-1}, somewhat different from the

value calculated based on the averaged GSAS lattice parameter fluctuations, and closer to the literature value and the AS800 value. This illustrates the value of ALPHA, as the averaged lattice values are somewhat incorrect due to the SiC overlap, thus using non-overlapping reflections gives a more precise value. The aspherism index for this Si_3N_4 was A = 0.032. SiC is cubic, so its thermal expansion is isotropic [14], thus ALPHA was not used. A lack of independent SiC reflections prevented this in any case.

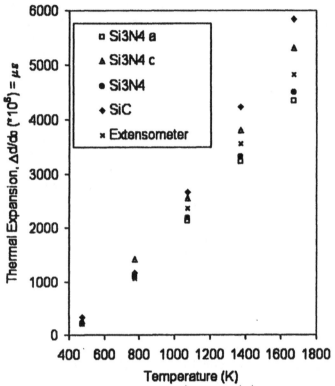

Figure 2. Thermal expansion of SiC_p-Si_3N_4 based on longitudinally measured lattice parameters. The CTE was calculated as $\alpha = 3.35 \times 10^{-6}$ K^{-1} for the polycrystalline average of Si_3N_4 and $\alpha = 4.33 \times 10^{-6}$ K^{-1} for SiC.

Diffraction Strain

AS800: Full-pattern data fits with GSAS provide the *a* and *c* lattice parameters for AS800 in these applied stress tests, averaged over the many peaks appearing in the patterns. Figure 3 shows the lattice strain in each direction, longitudinal and transverse, resulting from applied stress. Zero strain point in this figure is relative to the average lattice parameters of the 30MPa, 1375°C patterns. The other data points are also averages of the patterns collected for each stress, based on the linear behavior of the full data set. The Young's modulus (from a linear fit to the longitudinal data points) is 339GPa, while the Poisson's ratio is 0.32.

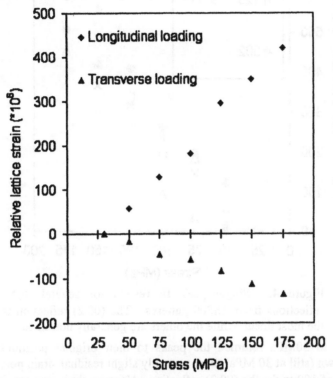

Figure 3. Average lattice strains resulting from applied stress for AS800. Linear fits give E = 339GPa and ν = 0.32.

Using single peak fits (of same reflections used for the CTE calculation) from both the transverse and longitudinal patterns, strains were calculated relative to the initial room temperature scan at 30MPa. The largest strains were realized due to thermal expansion, over 5000με for the (00·2) reflection. After reaching the test temperature and applying higher stresses, a maximum of 575με was realized, again for the (00·2) reflection, with significant variation of strains dependent upon the *hkl* of the reflection, shown for a few reflections in Figure 4.

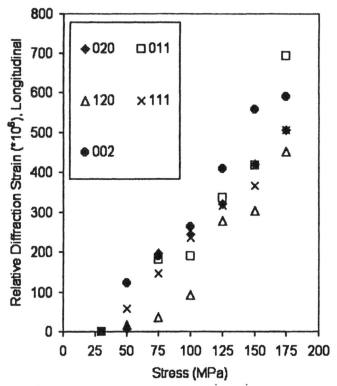

Figure 4. Single peak fit results for several Si₃N₄ reflections from AS800 patterns. The (00·2) reflection is the most linear, while the others are generally non-linear.

Unloading to 30MPa returned the peaks to their original positions. After cooling down (still at 30 MPa) there was only slight residual strain noted, with a maximum of 190με for the (00·2) reflection. Nor was there was any change in peak breadth, thus the deformation was almost entirely elastic. Calculation of the single crystal elastic constants of this material at this temperature was justified by this fact.

The 1375°C elastic stiffness tensor of AS800 was determined an elastic-plastic self-consistent (EPSC) polycrystal deformation model [16]. EPSC models have been demonstrated to predict accurately the diffraction elastic constants measured by neutron diffraction [16-19]. For these previous works, the only input was the material single crystal stiffnesses. For this research, a reverse procedure was used. The measured diffraction elastic constants and a least squares fitting routine were used to obtain the best fit for the single crystal stiffnesses. The starting point for the least squares refinement was the isotropic stiffness tensor calculated from the measured macroscopic Young's modulus and Poisson's ratio (Fig. 3). The multiple patterns from each stress were summed into a single pattern, and nine single peak fits were used to refine the stiffness tensor. This computation yielded the 1375°C stiffness tensor for AS800, with values shown in Table I.

Table I. Single crystal stiffness tensor values for AS800 Si_3N_4 at 1375°C compared with room temperature values for Si_3N_4.

	C_{11}	C_{33}	C_{44}	C_{66}	C_{12}	C_{13}
Present work	456 ± 31	311 ± 40	144 ± 11	149 ± 40	158 ± 40	238 ± 21
Ref [20]	433	574	108	119	195	127

Comparing the present values to those of Vogelgesang and Grimsditch [20] shows some agreement. As their test was both at room temperature *and* for a different grade of Si_3N_4, it is uncertain which is the source of the discrepancy. AS800 is known to have elongated grains (c-axis), which could account for the marked change in the C_{33} value. A true comparison will require a room temperature test of this material to obtain the room temperature stiffness tensor. Using the tensor to determine the macroscopic values for the Young's modulus and Poisson's ratio gives 313GPa and 0.31, respectively, very close to the values found from the average diffraction data. Honeywell reports these values at 1200°C as 293GPa and 0.28. The higher values obtained here are likely a result of this test being in vacuum, which will protect the sample from oxidation known to decrease mechanical properties. The large scatter for the various reflections in the SiC_p-Si_3N_4 sample prevented a similar stiffness tensor refinement.

SiC_p-Si_3N_4: Figure 5 shows the diffraction strains from both detector banks as a function of applied stress. The strains shown are relative to the lattice spacing measured for 30MPa at 1400°C. While not shown, the extensometer data exhibited the same behavior as the c-axis for the Si_3N_4, also not shown. Figure 6 shows the variation of strain for specific reflections. The erratic behavior is similar to that in Figure 4, but more pronounced. Thus is this behavior characteristic for strain applied to ISR-Si_3N_4 at high-temperature. While the scatter was significant for AS800, the stiffness tensor was still calculable. For the composite such calculation was not possible. Instead, and Eshelby inclusion model [21] was used to compare with the diffraction strain data. Literature values were used for the Young's moduli and Poisson's ratios for each phase. In the case of the silicon nitride values, the room temperature literature values [22] were multiplied by 0.85, to account for the softening of Si_3N_4 due to the higher temperature (estimated from Figure 2 in [23]) which

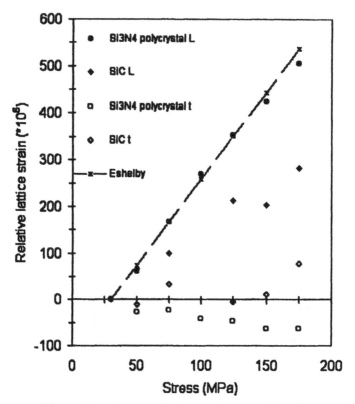

Figure 5. Average lattice strains resulting from applied stress for Si_3N_4-SiC_p at 1400°C. Linear fits give E = 281GPa and ν = 0.13 for Si_3N_4. L - longitudinal, t - transverse

seems a suitable value for both Si_3N_4-SiC_p composites and "pure" Si_3N_4. Using the volume fraction of the SiC particles (20vol% as determined with GSAS) and assuming a thermal misfit of 1000°C, the effect imposed by this particulate presence was calculated. The model calculations are shown in Figure 5 along with the GSAS diffraction strains. The model results coincide with the data for the Si_3N_4 polycrystalline average. This indicates that the model is successful for predicting the behavior of this system under stress *at high temperature*.

CONCLUSION

This investigation has demonstrated the capability of the SMARTS diffractometer to obtain microstructural information from tensile tests at high temperature in vacuum. The thermal expansion coefficients were measured for

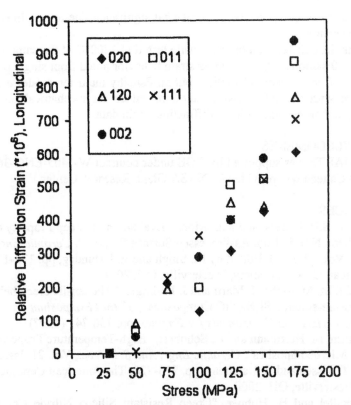

Figure 6. Single peak fit results for several Si₃N₄ reflections from SiC$_p$-Si₃N₄ patterns. All are non-linear, with 00·2 being most linear.

two different samples of ISR Si₃N₄, one monolithic, the other a composite with SiC particles. The anisotropic thermal expansion tensor components were computed using single peak fits of diffraction data. These gave CTEs similar to literature, but more accurately than using full pattern-derived lattice parameter values, especially in the composite sample which had significant diffraction peak overlap between the matrix and reinforcing phase.

These initial tests made use of imposed elastic deformation of tensile samples. The AS800 sample was mechanically loaded at 1375°C. Neutron diffraction data indicated deformation was mostly elastic, since, upon unloading and cooling, nearly all the deformation was fully recovered. Based on this, the high-temperature elastic tensor was calculated using single peak fits from high-temperature applied stress diffraction patterns. The tensor values are comparable to room temperature values, with softening in the c-direction. Error bars on the tensor components are somewhat significant due to non-linearity of specific reflections.

The SiC_p-Si_3N_4 sample was tensile tested at 1400°C at the same stresses as the AS800 sample. The non-linear strain behavior noted from single peak fits prevented stiffness tensor calculation, but an Eshelby inclusion model was used to compare with the data. Using literature values for mechanical properties yielded a good agreement with the diffraction strain data.

ACKNOWLEDGMENTS

SMARTS was supported by DOE under contract W-7405-ENG-36. The funding at Caltech was provided by NASA Glenn Research Center.

REFERENCES
1. H. T. Lin, S.B. Waters, and K.L. More, "Evaluation of Creep Property of AS800 Silicon Nitride from As-Processed Surface Regions", *Ceramic Engr. And Sci. Proc.*, Vol. 22, Iss. 3, Edited by M. Singh and E. Üstündag, pp. 175-182, The American Ceramic Society, Westerville, OH, 2001
2. C.-W. Li, M. M. Matic, J. Marti and J. Pollinger, "The tensile creep behavior of gas-pressure-sintered Si_3N_4-SiC Composites", *6th Int'l Symposium on Ceramic Materials and Components for Engines*, pp. 736-741 (1997)
3. H. Klemm, M. Herrmann and C. Schubert, "High-Temperature Properties of Si_3N_4/SiC Microcomposites", *Ceramic Engr. And Sci. Proc.*, Vol. 21, Iss. 3, Edited by T. Jessen and E. Üstündag, pp. 713-720, The American Ceramic Society, Westerville, OH, 2000
4. A. Rendtel and H. Hübner, "Creep Resistant Silicon Nitride Ceramics - Approaches of Microstructural Design", *Ceramic Engr. And Sci. Proc.*, Vol. 21, Iss. 4, Edited by T. Jessen and E. Üstündag, pp. 515-526, The American Ceramic Society, Westerville, OH, 2000
5. C. J. Gasdaska, "Tensile Creep in an *in Situ* Reinforced Silicon Nitride", *J. Am. Cer. Soc.*, 77 [9] 2408-2418 (1994)
6. C. W. Li and F. Reidinger, "Microstructure and tensile creep mechanisms of an *in situ* reinforced silicon nitride", *Acta Mater.*, 45 [1], 407-421 (1997)
7. H.M.A. Winand, A.F. Whitehouse, P.J. Withers, "An investigation of the isothermal creep response of Al-based composites by neutron diffraction", *Materials Science and Engineering*, A284, pp. 103-113 (2000)
8. M. R. Daymond, C. Lund, M.A.M. Bourke, and D. Dunand, "Elastic Phase-Strain Distribution in a Particulate-Reinforced Metal-Matrix Composite Deforming by Slip or Creep", *Metallurgical and Materials Transactions A*, 30A, pp. 2989-2997 (1999)
9. M.A.M. Bourke, D.C. Dunand, E. Ustundag, "SMARTS - a spectrometer for strain measurement in engineering materials", *Applied Physics A*, in press

2002

10. H. M. Rietveld, "A Profile Refinement Method for Nuclear and Magnetic Structures ", *J. Appl. Cryst.*, **2**, pp. 65-71 (1969)

11. A. C. Larson, R. B. von Dreele, *GSAS-General Structure Analysis System*, LAUR 86-748, Los Alamos National Laboratory, 1986

12. P. Villars and L.D. Calvert, *Pearson's Hanbook of Crystallographic Data for Intermetallic PhasesI*, Vol. 3, pg 2790, American Society for Metals, Metals Park, OH, 1985

13. S. M. Jessen and H. Küppers, "The Precision of Thermal-Expansion Tensors of Triclinic and Monoclinic Crystals", *J. Appl. Cryst.*, **24**, pp. 239-242 (1991)

14. D. Sands, *Vectors and Tensors in Crystallography*, pp. 138-9, Dover Publications Inc., New York, 1995

15. W. D. Callister, Jr., *Materials science and Engineering: An Introduction*, 3rd ed., pg. 768, John Wiley and Sons, Inc., New York, 1994

16. P. A. Turner, C. N. Tomé, *Acta Metall. Mater.*, **42**, 4143-4153 (1994)

17. B. Clausen, T. Lorentzen, and T. Leffers, *Acta mater.*, **46**, 3087-3098 (1998)

18. B. Clausen, T. Lorentzen, M. A. M. Bourke, and M. R. Daymond, *Mat. Sci. & Eng. A*, **259**, 17-24 (1998)

19. T. M. Holden, R. A. Holt, C. N. Tomé, *Mat. Sci. & Eng. A*, **282**, 131-136, (2000)

20. R. Vogelgesang and M. Grimsditch, "The elastic constants of single crystal β-Si$_3$N$_4$", *Appl. Phys. Lett.*, **76** [8], pp. 982-984 (2000)

21. T. Clyne and P. Withers, "An Introduction to Metal Matrix Composites", pp. 44-63, Cambridge University Press, 1993

22. D. Richerson, *Modern Ceramic Engineering: Properties, Processing and Use in Design*, 2nd ed., pp. 166-169, Marcel-Dekker, Inc., New York, 1992

23. T. Rouxel, J. Sangleboeuf, M. Huger, C. Gault, J. Besson, S. Testu, "Temperature dependence of Young's modulus in Si$_3$N$_4$-based ceramics: roles of sintering additives and of SiC particle content", *Acta Mat.*, **50**, pp1669-82 (2002)

Improved Properties

SIALON CERAMICS: PROCESSING, MICROSTRUCTURE AND PROPERTIES

Stefan Holzer[a], Bernd Huchler[b], Alwin Nagel[b], Michael J. Hoffmann[a]
[a]Institut für Keramik im Maschinenbau   [b]Fachhochschule Aalen
Universität Karlsruhe (TH)   Beethovenstr. 1
Haid-und-Neu-Straße 7   D-73430 Aalen
D-76131 Karlsruhe   Germany
Germany

ABSTRACT

Sialon ceramics with neodymia and ytterbia as sintering additive and with starting compositions of fixed n value (1.0) and varying m value (0.4 to 1.0) have been prepared. The amount of the rare-earth oxide exhibits a strong effect on sinterability, microstructure and mechanical properties. Ytterbium based sialons require excess Yb_2O_3 for gas pressure sintering. Dilatometer measurements reveal differences between the shrinkage curves. Increasing the amount of sintering aid shifts the microstructure of pure α-sialon ceramics from an equiaxed to an elongated grain morphology. This coincides with a clear increase of the fracture toughness. In mixed α/β-sialons the aspect ratio of the β-sialon phase rises. For two neodymium based sialons with different α:β-sialon ratios the 4-point-bending strength has been determined and their tribological behavior investigated.

INTRODUCTION

Sialons are a typical representative for a ceramic alloy. Depending on the application requirements the mechanical properties can be customized by altering parameters like the α:β-sialon ratio, the additive element or the sintering conditions. α-sialon is the harder modification [1] caused by additional cations on interstices in the structure which hinder the dislocation movement and increase the number of chemical bonds. The general formula $M_xSi_{12-m-n}Al_{m+n}O_nN_{16-n}$ (M = Li, Mg, Ca, Y, rare-earth elements like Nd, Yb) can be used to describe the composition of the α-sialon phase as well as to calculate the overall composition of a powder mixture. β-sialon crystals grow more needle-like and thus enable crack wake mechanisms leading to a higher fracture toughness. The α-sialon stability region and thus the α:β-sialon ratio can be determined by the thermal conditions [2, 3] and the choice of the additive element [4]. By controlling nucleation and growth one can change the microstructure from finegrained and equiaxed to large elongated grains [5]. Microstructural improvements by addition of different amounts of sintering aid in yttria based sialons have been achieved recently [6].

Si_3N_4 based ceramics are suited for mechanical and high temperature stresses and show a potential for tribological applications. Since the friction and wear behavior of silicon nitride and sialon strongly depends on the investigated system, a general statement is not possible. For instance, when testing silicon nitride in contact with gray iron and water lubrication, tribochemical reactions lead to a low friction coefficient and wear rate [7]. Without water the same system suffers from high wear rates [8]. An interesting tribological system is the sliding contact under lubrication with isooctane. It could enable the design of components for pumping gasoline under high pressure. So far, there is not much data available about this combination.

Limiting factors in the fabrication of ceramics are shaping and production costs. Hot-pressing confines the parts to simple shapes but this limitation can be overcome by isostatic pressing. To avoid high costs from pressures of up to 100 MPa, gas-pressure sintering with a maximum pressure of 10 MPa is chosen in this investigation. The role of an excess content of the sintering aid neodymia or ytterbia is analyzed in this paper.

EXPERIMENTAL PROCEDURE

Si_3N_4 (UBE, SN-E10), AlN (H. C. Starck, grade C), Al_2O_3 (Alcoa, CTC 20), Nd_2O_3 (Alfa Aesar, Neodymium (III) Oxide REacton), and Yb_2O_3 (Shin-Etsu, Ytterbium Oxide) have been used as starting powders. The amount of oxygen in silicon and aluminum nitride has been taken into account. After attrition milling for 4 h with silicon nitride milling balls in isopropanol the slurries were dried in a rotary evaporator and sieved. Rectangular specimens for the characterization of microstructure and mechanical properties as well as tribological experiments were uniaxially and subsequently cold isostatically pressed at 400 MPa pressure. The tribological counterparts had a cylindrical shape and were consolidated by isostatical pressing of the powder mixture in a rubber preform. Smaller pieces were sintered in a BN crucible under nitrogen atmosphere in a hot isostatic press with a graphite resistant heater. The heating rate was 10 K/min to the maximum sintering temperature of 1800 °C. The pressure has been kept below 1.0 MPa up to 1800 °C and then increased to 10 MPa. After a dwell time of 60 min the furnace was switched off. Because of the retarded temperature increase in big samples cylinders and large plates were sintered differently. The maximum temperature was 1830 °C at which the pressure was 1 MPa of nitrogen. After a dwell time of 15 min argon pressure of 10 MPa was applied for another 45 min. The densification behavior has been determined with a dilatometer inside the furnace. Relative density was calculated from raw powder densities in consideration of the relative amount in the powder mixture.

Samples for microstructural observation and indentation measurements were cut in half and polished. The scanning electron microscope (Stereoscan 440, Leica UK) was used in backscattered electron mode. X-ray diffraction data were utilized for the calculation of α/β-sialon ratios by using the formula set up by Gazzara and Messier [9]. Hardness and fracture toughness were measured with a Akashi hardness tester (AVK-C1) equipped with a Vickers indentor and a load of 10 kg. The equation for the calculation of the fracture toughness was proposed by Anstis et al. [10]. Niihara et al. [11] refined it for the case of Palmqvist cracks to be

$$K_{Ic} = 0.018 \ H \ a^{0.5} \ (E/H)^{0,4} \ (c/a - 1)^{-0.5} \tag{1}$$

where E is the Young's modulus, H the Vickers hardness, a the half length of the indentation diagonal, and c crack length. The value of E was assumed to be 320 GPa which is in good aggreement with the measurement of two sialon ceramics used in this work. The resonance method resulted in 317 GPa and 320 GPa, respectively, for neodymia based sialons with different α:β-sialon ratios. The strength has been determined in a 4-point-bending test with specimens of 45 x 4 x 3 mm³ and a span width of 20 mm, respectively. Weibull statistics were calculated from data from 30 specimens.

Friction and wear tests were conducted on an SRV III tester (Optimol Germany) using an oscillating sliding movement of a ceramic cylinder against a ceramic disc (Fig. 1). Before each test, the specimens were cleaned with ethanol. The oscillatory sliding frequency was 20 Hz with an amplitude of 2 mm. Temperature was controlled at 20 ± 5 °C using a water cooled peltier heating system. The testing time was 30 min, corresponding to 288 m sliding distance. The normal force of the cylinder versus the disc was 60 N for all tests, resulting in an initial hertzian surface pressure according to Table I, calculated with equation (2). Isooctane was applied to simulate a self lubricating bearing. The isooctane was added onto the specimens continuously with 0.3 ml/min. The friction coefficient f was recorded online by the SRV tester. Wear was recorded offline at the middle part of the wear trace with a profilometer. In addition, microstructural investigations were conducted.

$$p_0 = 0,418 \sqrt{\frac{F \cdot E}{r \cdot h}} \tag{2}$$

Equation (2): Calculation of the initial surface pressure p_0 for cylinder-on-disc geometry with F = load (60 N), E = modulus of elasticity (MPa, see Table I), r = radius of cylinder (5 mm), and h = height of cylinder (22 mm).

For each material type both the cylinder and the disc were made of identical material and with the same final grinding conditions. Different grit sizes were used to investigate the influence of the surface grinding conditions (Table I). The Si_3N_4 and the Al_2O_3 ceramic are commercial materials both with a Vickers-hardness of 1650 under a load of 0.5 kg.

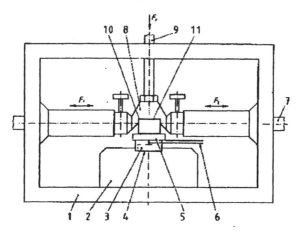

Fig. 1: Test chamber (schematic). 1 test chamber, 2 heating / cooling block, 3 load cell, 4 thermocouple, 5 disc, 6 lubricant supply, 7 oscillating rods, 8 cylinder bracket, 9 charging unit, 10 cylinder.

Table I. Surface finishing and calculated initial surface pressure.

Material α:β ratio (sialon)	Si_3N_4		Sialon 55:45	88:12	Al_2O_3
Finishing grit size	D25	D91	D91	D91	D46
R_a [μm]	0.19	0.44	0.36	0.42	0.75
R_z [μm]	2.08	3.63	2.30	2.26	5.71
E-Modulus [GPa]	305	305	320	320	380
Surface pressure [MPa]	170	170	175	175	190

RESULTS AND DISCUSSION
Densification Behavior

Gas pressure sintering of sialon ceramics brings along difficulties in obtaining complete densification. In the present work the role of different starting compositions and additive elements has been investigated. According to the general sialon formula, the m value has been varied from 0.4 to 1.0 with a fixed n value of 1.0. Because they differ significantly in their cation radii, neodymium and ytterbium have been chosen as stabilizer cations. The third parameter is the amount of additive in each respective composition. Starting from a powder mixture which corresponds to a composition within the sialon plane, the amount of neodymia or ytterbia was increased relative to the initial amount. In the case of Nd-sialon with m = 0.4, for instance, an excess of 40 % means 5,41 wt.-% Nd_2O_3 in the powder mixture instead of 3,86 wt.-%.

Ytterbia based sialons without an excess of Yb_2O_3 are hard to solidify. Relative densities between 83 % for m = 0.4 and 62 % for m = 1.0 have been achieved (Fig. 2). The continuously decreasing density can be connected to the increase in the α-sialon content from 45 to 100 % (Fig. 3). Thus, α-sialon rich compositions are more difficult to densify. The sintering behavior changes completely when small amounts of ytterbia are added. With 20 or 40 % more Yb_2O_3 in the powder mixture all samples show relative densities exceeding 97 %, independent of the m value.

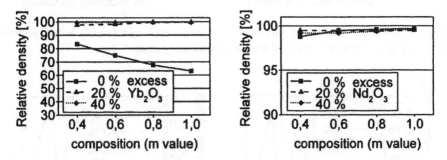

Fig. 2: Relative sintering densities of sialons based on ytterbia (left) and neodymia (right) as a function of the m value and the excess additive amount.

As can be seen in Fig. 2, neodymia containing sialons can be sintered much easier. Relative densities of at least 98 % have been obtained in all cases. Possible explanations for the different effects of neodymium and ytterbium on the solidification process are an oxygen pick-up and the wetting behavior. During powder processing in an open milling container additional oxygen may be absorbed by the slurry. Silicon and aluminum nitride then react to the corresponding oxides which enhance solidification due to a higher amount of liquid phase. An increase in the oxygen content by 1-2 wt.-% would be enough to achieve this effect. Comparing the α-sialon content of sialons with neodymia and ytterbia (Fig. 3) fortifies this assumption. Although the stability regions of α-sialon with neodymium and ytterbium differ significantly [4], the specimens investigated in this work show similar α-sialon contents. This can be attributed to a higher oxygen absorption of Nd containing samples which promotes the formation of the α-sialon modification. The second possible explanation is given by Menon and Chen [12, 13] which proved a different wetting behavior for both oxides. Smaller cations like Yb wet and react first with AlN, whereas larger cations like Nd prefer Si_3N_4.

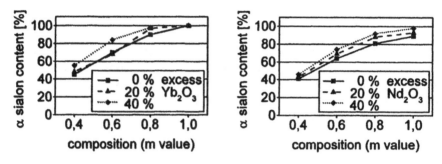

Fig. 3: α-sialon content of ytterbia (left) and neodymia (right) based sialons, depending on the m value and the excess additive content.

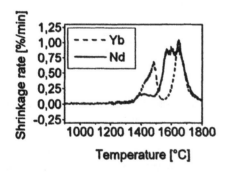

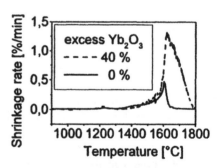

Fig. 4: Densification behavior of α/β-sialons (m=0.4) with Nd or Yb as additive (40 % excess additive).

Fig. 5: Densification behavior of Yb-α-sialons (m=1.0) with different additive contents.

The progress during sintering has been monitored by dilatometer measurements within the furnace and fits well the results on relative densities. Fig. 4 shows the dependency of solidification on the additive element. In both cases the maximum shrinkage rate is rather the same. But with neodymium the main peak has its rise some 60 °C below the corresponding peak of the ytterbium system and the shrinkage holds on for a longer time. The dilatometer results for ytterbia based α-sialons at different excess amounts of additive are in good agreement with the measured final densities, too (Fig. 5). In the beginning, both specimens show a similar curve progression. Liquid phase is formed between 1300 and 1400 °C and the solidification starts. Crystallization of α-sialon consumes the ytterbium cation and without an excess amount of Yb_2O_3 all ytterbium is spent before the densification is completed.

Microstructural development

The poor densification of ytterbia based sialons without an excess of a sintering aid leaves up to 40 % pores in the specimen. Even compositions containing higher amounts of β-sialon reach only some 80 % relative density. The fine-grained microstructure consists of crystals of both sialon modifications (Fig. 6) with α-sialon being the brighter one in the scanning electron micrograph. The grain boundary phase can be seen at the triple points as white spots. Black areas in the left micrograph indicate pores. Small amounts of extra Yb_2O_3 are enough to complete the solidification process and have no effect on the grain morphology.

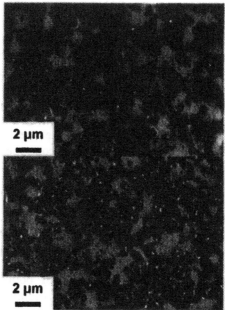

Fig. 6: Yb-α/β-sialon (m = 0.4) without an excess additive content and insufficient densification (left) and with 20 % excess Yb$_2$O$_3$ and full densification (right).

Higher amounts of additional ytterbia lead to a larger portion of liquid phase during sintering and change the grain growth of the β-sialon phase significantly (Fig. 7). "Free" space for growth is considered to be the promoting factor [14]. Crystals with aspect ratios of up to 10 can be found. As customary, the α-sialon grows equiaxed. The amount of grain boundary phase increases too, but is still not higher than in silicon nitride ceramics. As illustrated in the diagrams of Fig. 3 the relative amount of the α-sialon phase increases with higher oxide contents. The microstructure of pure α-sialon compositions is also dependent on the sintering aid content. A higher volume fraction of liquid phase modifies the growth conditions and the microstructure changes from equiaxed α-sialon grains to a bimodal distribution with smaller crystals and a fraction of larger, needle-like grains (Fig. 8). This gives the possibility of an in-situ reinforcement of sialons ceramics that enables crack wake mechanisms as known from silicon nitride ceramics [15]. The second presupposition for increasing the fracture toughness is a weak grain boundary phase [16].

Fig. 7: Yb-α/β-sialon (m = 0.4) with 40 % excess Yb_2O_3 causing an increase of the aspect ratio of the β-sialon phase.

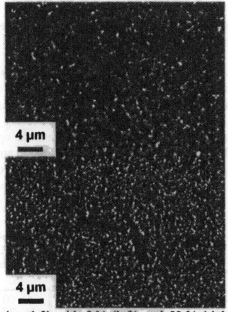

Fig. 8: Yb-α-sialon (m =1.0) with 5 % (left) and 50 % (right) more Yb_2O_3. The higher additive content promotes a bimodal microstructure with small grains and larger ones with increased aspect ratio.

Mechanical properties

The elongated grain growth in the pure α-sialon specimen with excess additive content has a clear effect on the fracture toughness. We achieved an increase from 5.3 MPam$^{1/2}$ for the sample with equiaxed microstructure to 6.4 MPam$^{1/2}$ for the sample with needle-like α-sialon grains (Fig. 9). Nevertheless, this advantage comes along with a reduction of the Vickers-hardness from 2030 to 1900 caused by the higher amount of intergranular glassy phase.

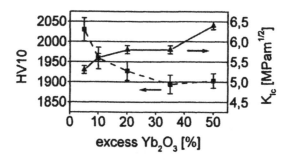

Fig. 9: Hardness and indentation fracture toughness of Yb-α-sialon as a function of the excess additive content.

In Fig. 10 the dependency of Vickers-hardness and indentation fracture toughness is plotted versus composition and excess additive content for neodymia based sialons. Due to a higher α-sialon fraction the hardness rises with increasing m value from 1747 to 1864 without excess Nd_2O_3. Decreasing the amount of elongated β-sialon grains upon increasing the m value has the opposite effect on the fracture toughness. The measured values for both hardness and fracture toughness at a fixed m value are rather close and no explicit influence of different sintering aid contents on the mechanical properties of neodymia based α/β-sialons can be found.

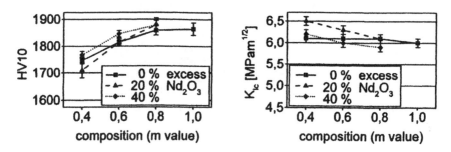

Fig. 10: Hardness and indentation fracture toughness of Nd-α/β-sialon depending on the m value and the excess rate.

For two neodymia based sialons with different α:β-sialon ratios the strength has been measured under 4-point-bending (Fig. 11). A relatively high strength of 863 MPa has been achieved for the α-sialon rich material with a ratio of 88:12 compared to 748 MPa for the composition with a ratio of 55:45. The Weibull parameter had values of 11.6 and 10.4, respectively. The higher value of the 4-point-bending strength can be attributed to two parameters. First, the higher amount of α-sialon crystals which are more equiaxed and smaller than the β-sialon grains. Second, a higher amount of grain boundary phase which was detected in the micrograph.

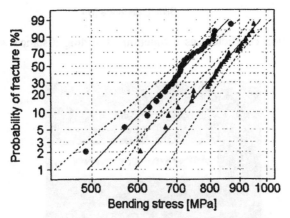

Fig. 11: 4-point-bending strength of neodymia based sialons with α:β = 55:45 (●) and α:β = 88:12 (▲).

Tribological Behaviour

First measurements of oscillating sliding tests were performed. The friction coefficient f is shown in Fig. 12 as a function of sliding distance for the different materials. Alumina shows the highest initial friction coefficient of f = 0.74 for all materials; during sliding f decreases to 0.55 with some instable behaviour. The silicon nitride and sialon grades show an initial friction coefficient f = 0.25 with increasing values but remaining lower than alumina. Silicon nitride with rougher surface D91 shows significantly higher instabilities compared to the smoother finishing D25. Both silicon nitride surfaces, D91 and D25, show the same increase of friction to f = 0.40 at the end of the sliding distance. The lowest friction coefficient was observed for the sialon grades. A final friction coefficient f = 0.3 was found for both different sialons. Wear traces are observable, but wear depth could not be measured because the wear was smaller than the surface roughness (Fig. 13).

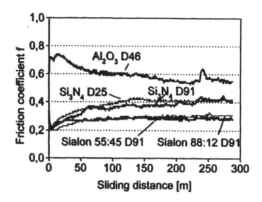

Fig. 12: Friction coefficient as a function of sliding distance with isooctane lubrication.

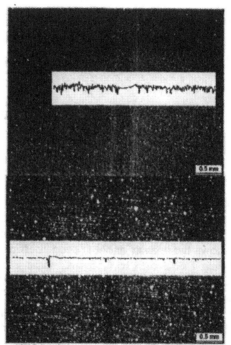

Fig. 13: Wear path and surface profile of alumina (left) and silicon nitride D25 (right). Wear depth is too small to be measured. Al_2O_3 shows a polished surface in the wear path.

Silicon-Based Structural Ceramics

Macroscopic structure of the disc wear traces are compared in Table II. The lowest wear path width was 310 μm for the neodymium based sialon with α:β = 88:12. Nevertheless, all specimens showed very low wear.

Table II: Width of wear band and microscopic structure of wear path.

Ceramic α:β ratio (sialon)	Si₃N₄ D25	Si₃N₄ D91	Sialon 55:45	Sialon 88:12	Al₂O₃
Width of wear band [μm]	340	410	410	310	430

SUMMARY

The densification behavior, the microstructural development and the resulting mechanical properties for neodymia and ytterbia based sialons with a fixed n value of 1.0 and m values between 0.4 and 1.0 have been characterized as a function of the additive content. For sialons with ytterbia as sintering aid it is necessary to add an excess amount of Yb_2O_3 for complete solidification. Higher rare-earth oxide contents caused an elongated grain growth of the β-sialon phase in mixed α/β-sialon samples and of the α-sialon phase in pure α-sialon ceramics. Elongated grains in pure α-sialons cause an increase in fracture toughness from 5.3 MPam$^{1/2}$ to 6.4 MPam$^{1/2}$. 4-point-bending strength and the response on tribological stress have been examined for two neodymium based sialons with different α:β-sialon ratios. A relatively high strength has been found for the material with a ratio of α:β = 88:12 and a higher fraction of grain boundary phase. Cylinder-on-disc testing demonstrated the high potential of sialons under isooctane lubrication. Sialons show the lowest friction coefficient in contact with isooctane compared with silicon nitride and alumina materials.

FUTURE PERSPECTIVES

Silicon nitride based ceramics have been explored since more than 4 decades and an enormous progress has been made to understand the complexity between powder properties, microstructural development and the resulting properties. Despite these significant improvements in tailoring properties, the expected breakthrough in the field of applications could not be realized. There are no real mass applications for service temperatures above 1100°C. The use of silicon-based materials as gas turbine components is limited due to the insufficient corrosion resistance at high water vapour pressures. Another main obstacles for a wider range of applications are the high costs especially in the case of small series. Today, profitable products are restricted to cutting tools, ball bearings, and some industrial parts. The experiences over the past decade shows that ceramics can only substitute metal components if there are clear advantages in the performance of the whole system which compensates for the higher prize of the ceramic part. New markets especially in the area of wear parts can only be penetrated if researchers are able to demonstrate the high potential of silicon nitride based ceramics components in a more complex system. This requires a tailoring of the materials properties for specific applications under consideration of the environmental boundary conditions as well as a special component design. Therefore, we need a much closer collaboration between material scientists and mechanical engineers to develop whole systems instead of improving single properties such as strength and toughness.

ACKNOWLEDGEMENT

This work was supported by the Deutsche Forschungsgemeinschaft (German Research Foundation) within project B1 of the center of excellence SFB 483 "High performance sliding and friction systems based on advanced ceramics".

REFERENCES

[1] T. Ekström and M. Nygren, "SiAlON Ceramics," J. Am. Cer. Soc. 75 [2] 259-276 (1992).

[2] M. Mitomo and A. Ishida, "Stability of α-Sialons in Low Temperature Annealing," J. Eur. Cer. Soc. 19 7-15 (1999).

[3] N. Camuşcu, D. P. Thompson and H. Mandal, "Effect of Starting Composition, Type of Rare Earth Sintering Additive and Amount of Liquid Phase on α↔β Sialon Transformation," J. Eur. Cer. Soc. 17 599-613 (1997).

[4] S. Holzer, H. Geßwein and M. J. Hoffmann, "Phase Relationships in Neodymia and Ytterbia Containing SiAlONs," Key Eng. Mater. 237 43-48 (2003).

[5] Z.-H. Xie, M. Hoffman and Y.-B. Cheng, "Microstructural Tailoring and Characterization of a Calcium α-SiAlON Composition," J. Am. Cer. Soc. 85 [4] 812-818 (2002).

[6]S. Kurama, M. Herrmann and H. Mandal, "The effect of processing conditions, amount of additives and composition on the microstructures and mechanical properties of α-SiAlON ceramics," J. Eur. Cer. Soc. **22** 109-119 (2002).

[7]Y.-M. Gao, L. Fang, J.-Y. Su et al., "Investigation on the components and the formation of a tribochemical film in the Si_3N_4-gray iron sliding pair lubricated with distilled water," Wear **206** 87-93 (1997).

[8]L. Fang, Y. Gao, L. Zhou et al., "Unlubricated sliding wear of ceramics against graphitized cast irons," Wear **171** 129-134 (1994).

[9]C. P. Gazzara and D. R. Messier, "Determination of Phase Content of Si_3N_4 by X-Ray Diffraction Analysis," Am. Cer. Soc. Bull. **56** [9] 777-780 (1977).

[10]G. R. Anstis, P. Chantikul, B. R. Lawn and D. B. Marshall, "A Critical Evaluation of Indentation Techniques for Measuring Fracture Toughness: I, Direct Crack Measurements," J. Am. Cer. Soc. **64** [9] 533-538 (1981).

[11]K. Niihara, R. Morena and D. P. H. Hasselman, " Evaluation of K_{Ic} of brittle solids by the indentation method with low crack-to-indent ratios," J. Mater. Sci. Letters **1** 13-16 (1982).

[12]M. Menon and I-W. Chen, "Reaction Densification of α'-SiAlON: I, Wetting Behavior and Acid-Base Reactions," J. Am. Cer. Soc. **78** [3] 545-552 (1995).

[13]M. Menon and I-W. Chen, "Reaction Densification of α'-SiAlON: II, Densification Behavior," J. Am. Cer. Soc. **78** [3] 553-559 (1995).

[14]M. J. Hoffmann and G. Petzow, „ Tailored microstructures of silicon nitride ceramics," Pure & Appl. Chem. **66** [9] 1807-1814 (1994).

[15]P. F. Becher, E. Y. Sun, K. P. Plucknett et al., "Microstructural Design of Silicon Nitride with Improved Fracture Toughness: I, Effects of Grain Shape and Size," J. Am. Cer. Soc. **81** [11] 2821-2830 (1998).

[16]E. Y. Sun, P. F. Becher, K. P. Plucknett et al., "Microstructural Design of Silicon Nitride with Improved Fracture Toughness: II, Effects of Yttria and Alumina Additives," J. Am. Cer. Soc. **81** [11] 2831-2840 (1998).

FRACTURE BEHAVIOR OF POROUS Si_3N_4 CERAMICS WITH RANDOM AND ALIGNED MICROSTRUCTURE

Jian-Feng Yang, Naoki Kondo and Tatsuki Ohji
Synergy Materials Research Center, National Institute of Advanced Industrial
Science and Technology (AIST), Nagoya 463-8687, Japan.
Yoshiaki Inagaki
Synergy Ceramics Laboratory, Fine Ceramics Research Association, Nagoya 463-8687, Japan.

ABSTRACT

Porous Si_3N_4 ceramics with porosity content from 0 to 30% was fabricated by three techniques: (1) partial hot-pressing (PHP), (2) sinter-forging technique, and (3) tape casting of β-Si_3N_4 whisker. With PHP method, the porosity was controlled by the amount of powder addition and the microstructure could be similar for samples with different porosity. Applying the PHP to sintering forging, porous silicon nitride with aligned fibrous grains and controlled porosity was fabricated. Using tape-casting technique, a porous silicon nitride with aligned fibrous β-Si_3N_4 grains was obtained. Porosity dependence of Young's modulus, flexural strength, and fracture toughness (K_{IC}), were investigated. All of these properties decreased with increasing porosity. However, the strain tolerance (fracture strength/Young's modulus) and the critical energy release rate increased with increasing porosity, but decreased with further increasing porosity. The porous Si_3N_4 ceramics, containing a highly aligned whisker or fibrous β-Si_3N_4 grains and pores, exhibited an excellent fracture resistance and strength.

INTRODUCTION

The outstanding properties of silicon nitride (i.e. high strength and hardness, low thermal expansion, and super chemical durability) make these ceramics of interest for application as high-temperature structural components. Because these ceramics are difficult to machine, it is desirable to make a ceramic component with suitable size and simple shape. In industrial applications, most of the ceramic components are assembled with the metallic ones to utilize the individual outstanding properties. Due to the mismatch in thermal expansion coefficient between the ceramics and metals, the thermal stress arises. Thus, the ceramic components are easily fractured due to the presence of residual tensile stress. To overcome this problem, it is strongly desirable to reduce the Young's modulus of the ceramics without sacrificing the strength. For example, if the Young's modulus decreases by half, the thermal stress can be reduced by half as well, and strain tolerance can be enhanced.[1] Usually, the modulus can be decrease by half with 20-30% porosity.[2]

One of the approaches to decrease the Young's modulus is to introduce a second phase with a low Young's modulus such as BN [3,4] and carbon; however, these composites have low sinterability and oxidation resistance. The low sinterability restrains the sintering of materials, and the low oxidation resistance limits the application of material at high temperature and oxidizing environments. An alternative method is to introduce pores, forming porous ceramics. Besides the low Young's modulus, porous ceramics have many advantages such as good damage tolerance, lightweight, and low cost.

Many approaches have been attempted to introduce the pores in the ceramic microstructure. Porous Si_3N_4 ceramics can be fabricated by either using a removable substance [5,6] or controlled sintering of Si_3N_4 powder with low sintering additive. [7,8] Using the removable substance such as starch, the pore size is dependent on the size of the substance, so it is difficult to obtain very fine pores.[5] Alternatively, partial sintering is a suitable process to obtain porous ceramics with refined pores.[7,8] However, controlling of different parameters such as grain boundary composition, grain size, and final porosity simultaneously is difficult to achieve.

This paper summarizes the research results on the porous Si_3N_4 ceramics with <40% porosity. The purpose of development of these materials was to achieve above goal of decreasing the Young's modulus. Various techniques were developed to make the porosity, microstructure, and mechanical properties controllable. The following techniques had been developed: (1) partial hot-pressing (PHP);[9] (2) sinter-forging technique;[10] and (3) tape casting of β-Si_3N_4

whisker.[11] Samples with porosity ranged from ~0 to 30% were prepared to investigate the influence of porosity content on the mechanical properties of the porous Si_3N_4 ceramics. Densification, microstructure, and mechanical properties, including Young's modulus, flexural strength, fracture toughness, strain tolerance, and fracture energy as a function of porosity, were investigated.

EXPERIMENTAL PROCEDURE

1. Powder mixture

During the preparation of the porous Si_3N_4 ceramics, the densification was not desirable, so that the selection of sintering additives was important. Table 1 showed the composition of starting powders for each technique. In order to reduce sinterability, 5 wt% Yb_2O_3 or 5 wt% Y_2O_3 was used as the sintering additive for PHP and sintering-forging, respectively. The addition of Al_2O_3 in type casting is to enhance its sinterability. The powder mixture was wet-milled in methanol for 24 h, followed by dry and sieve through a 150 μm screen.

2. Preparation of sintering body

(1) PHP: The powder mixture was weighed and packed into the hot-pressing mold with an area of 42 mm × 45 mm. The PHP uses a specially designed mold where the sum of top and bottom punch lengths is shorter than the length of mold.[9] Thus, the pressure can no longer be applied onto the powder compact when the punches meet the ends of the mold. The difference between these two lengths becomes the thickness of the sample. The density can then be simply and precisely determined by the amount of starting powder. Sintering was conducted in a graphite resistance furnace (Model No. FVPHP-R-10, Fujidempa Kogyo Co., Ltd., Osaka, Japan), at a temperature of 1800°C using nitrogen atmosphere. The heating rate was 10-15 °C/min, and holding time was 2 h.

(2) Sinter-forging: The powder was compacted with a uniaxial pressure of 2.5 MPa in a graphite die with a base of 45 mm x 45 mm. Partial sinter forging was conducted using a hot press furnace. The graphite die was heated to 1850°C at a rate of 10°C / min. After soaking for 30 min, an uniaxial mechanical pressure of 6000 kg (approximately 30 MPa) was applied for 150 min (therefore, total soaking time at 1850°C was 180 min). Piston transfer was restricted to control the density of the sintered specimen, as the PHP.

(3) Tape casting: The green sheets, which were formed via a tape-casting method, were stacked and bonded under pressure. Sintering was performed at 1850°C under a nitrogen pressure of 1 MPa. The obtained porous Si_3N_4 (hereafter called PSN) was compared with a reference dense Si_3N_4 material (hereafter called RSN).

3. Characterization

The obtained sintering bodies were machined into testing bars with the dimensions of 3 mm x 4 mm x 42 mm (JIS R1601) for measurement of relative density, Young's modulus, strength, and fracture toughness. Each surface was finished with a 600-meshed diamond wheel. The bulk density and open porosity were measured by the Archimedes displacement method.

The Young's modulus, E (GPa), was measured using an ultrasonic technique according JIS 1602. Three-point bend testing (span of 30 mm) was performed to determine the flexural strength using a testing machine at a displacement speed of 0.5 mm/min. Before testing, the tensile edges on the tensile surface were beveled. Strain tolerance was determined by dividing the flexural strength by the Young's modulus.[1] Critical stress intensity factor, or fracture toughness, K_{IC}, was measured by the single-edged-precracked beam (SEPB) method according to JIS R1607. The strength and fracture toughness data were averaged from six measurements. With knowledge of E and K_{IC}, critical energy release rate, G_C, was determined by the following equation.[12]

$$G_C = \frac{K_{IC}^2 (1 - v^2)}{E} \tag{1}$$

Where v is the Poisson's ratio, assumed to be 0.25. The microstructure and fracture surfaces of the bending test were characterized by scanning electron microscopy (SEM; Model S-5000, Hitachi, Ltd., Tokyo, Japan).

For the CNB (Chevron Notch Beam) testing, the shape of the ligament was a regular triangle with an edge length of 3 mm, and the initial crack length (a_0) was 1.4 mm. The width of the chevron notch was 0.1 mm. The true load-displacement (L-D) curve was determined by subtracting the compliance of the testing machine and the fixture, which was obtained in advance by an independent calibration, from the experimentally observed curve.

Table 1: Composition of starting powders

Process	Si_3N_4 type	Si_3N_4 (wt%)	Y_2O_3 (wt%) [2]	Al_2O_3 (wt%) [4]
PHP	E10[1]	95	5 (Yb_2O_3) [3]	0
Sinter-forging	E10	95	5	0
Tape casting	β-whisker	93	5	2

1. E10, Ube Industries. Ltd., Tokyo, Japan, α ratio >95.5%, particle size 0.55 μm
2. UF, Shin-Etsu Chemical Co. Ltd., Tokyo, Japan
3. 99.9% purity, High Purity Chemical Co., Tokyo, Japan

RESULTS AND DISCUSSION

1. Mechanical properties of samples by PHP

In this study, five Si_3N_4 samples with a various porosity of 0-30% were prepared by the PHP method.

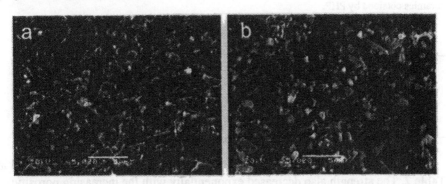

Figure 1 Microstructure of Si_3N_4 ceramic made by PHP. The porosity were (a) 0.4%, (b) 23.3%.

Microstructure: The microstructures of these materials are given in Figure 1 for the samples with porosity 0 and 23.3%. It showed that the grain morphology, including average grain size (average 0.5 μm) and aspect ratio (5-7) of the materials kept almost unchanged, as the sintering process was exactly the same. For all the specimens, the β-Si_3N_4 phase grain growth achieved after reaching the expected density and the interaction of the fibrous β-Si_3N_4 grains could be beneficial to the mechanical properties of the resultant specimens. It was also evident that the samples exhibited a fine pore distribution among the fibrous Si_3N_4 grains.

Young's modulus: Elastic modulus E is shown in Table 2. With increasing in porosity, the Young's modulus decreased obviously following an exponent relationship as usual:[13]

$$E(P) = E_0Exp(-\alpha P) = 328Exp(-3.249P) \qquad (2)$$

Where E_0, α and P are the Young's modulus at $P = 0$, the slope of the property dependence over the linear range and porosity, respectively. This equation is the most used among the typical equations for the porosity dependence of elastic modulus, primarily for equiaxial pore.[13] The microstructure in Figure 1 also demonstrates that the pore in the samples is equiaxial, so that the above relationship was suitable. On the other hand, as the Young's modulus is only porosity dependence but not sintering process dependence,[14] the elastic modulus can be taken as a reference for the porosity dependence of other mechanical properties.

Table 2. Porosity, Young's modulus, flexural strength, and strain to failure of the porous Si₃N₄ ceramics obtained by PHP.

Porosity	Young's modulus (GPa)	Flexural strength (MPa)	Fracture toughness (MPa·m$^{1/2}$)	Strain to failure ($\times 10^{-3}$)
0.004	317	1078±34	6.32	3.40
0.084	255	901±40	5.72	3.53
0.161	198	839±39	5.30	4.23
0.233	154	704±40	4.44	4.55
0.299	122	540±37	3.28	4.41

Flexural strength: The flexural strength as a function of porosity is shown in Table 2. The strength also decreased exponentially with the increasing porosity:

$$\sigma(P) = \sigma_0 Exp(-\alpha P) = 1113 Exp(-2.186P) \qquad (3)$$

Where σ_0 is the flexural strength at P=0. Comparing the value of α in Equation 2 and 3, the strength decreased moderately with the increasing porosity, comparing with the Young's modulus due to a low α value. This moderate decrease in strength resulted from the fibrous β-Si₃N₄ grain in the microstructure, and gave rise to an increase in strain to failure as shown in Table 2.

Fracture toughness: The fracture toughness is also illustrated in Table 2. The fracture toughness decreased with the porosity following a multinomial relationship:

$$K_{IC}(P) = K_{IC0}(1 - aP - bP^2) = 6.26(1 - 0.4085P - 3.85P^2) \qquad (4)$$

Where σ_0 is the fracture toughness at $P = 0$, and a and b constant. The standard error was 0.9959.

Table 3. Critical strain energy release rate, G_C, of the porous Si₃N₄ ceramics obtained by PHP.

Relative density (%)	0.004	0.084	0.161	0.233	0.299
G_C (J/m^2)	126	128	141	128	88

Critical strain energy release rate: The critical strain energy release rate, G_C, was obtained from K_{IC} and Young's modulus according to Equation (1), and shown in Table 3. The G_C remained unchanged when the porosity increased to 8%, but increased to a peak value when the porosity is about 16%. The porous

Si_3N_4 with 16% porosity exhibited a G_C value of about 140 J/m^2, which is 10% higher than that of dense one.

2. Mechanical properties of samples by sinter-forging

After the partial sinter forging, the porosity was controlled to be around 25%. The fabricated specimen showed the relative density of 76%, which contained 2.2 and 21.8% of closed and open pores, respectively. Microstructure of the specimen is shown in Figure 2. This micrograph was taken from the fractured top plane. Pressing direction in the side plane was vertical. The fibrous grains appear to be randomly oriented in the top plane, but somewhat aligned perpendicularly to the pressing direction in the side plane. The microstructure also shows protruding fibrous grains as well as holes or hollows. The porous microstructure with two-dimensionally oriented β-silicon nitride grains were successfully formed by the partial-sinter-forging technique.

Figure 2 Microstructure of the fabricated specimen. Some of the typical protruding fibrous grains are noted by arrows.

In this study, mechanical pressure was applied after grain elongation. The relative density of the green compact in the graphite die was roughly estimated as approximately 50%; therefore, height reduction of approximately 34% could be applied by the partial sinter forging, resulting in the anisotropic microstructure. From the previous studies,[15,16] the larger the silicon nitride deformed, the higher the anisotropy would be. Therefore, a material with lower porosity and higher anisotropy, or with higher porosity and lower anisotropy, can be obtained by choosing the sinter-forging condition.

The elastic modulus and strength of the fabricated specimen are summarized in Table 4. Elastic modulus was measured from top and side planes, respectively. The elastic modulus was 150 and 183 GPa, for the top and side planes, respectively. The top and side plane are perpendicular and parallel to the pressing direction, respectively. Elastic anisotropy of β-silicon nitride grain was measured by,[17] i.e., the elastic is 450 – 540 GPa in the direction parallel to the c axis, and is 280 – 310 GPa perpendicular to it. Due to the Si_3N_4 whisker exhibits a strong anisotropy as measured by Hay *et al.*, the elastic anisotropy in present specimens should originate from the anisotropic microstructure as most the c-axis perpendicular to the pressing direction.

Table 4. Characteristics of the fabricated specimen.

Relative density	Open porosity	Closed porosity	Elastic modulus(GPa)	Strength (MPa)
0.76	0.218	0.022	150 (top) 183 (side)	778 ± 17

The specimen exhibited relatively high bending strength of 778 MPa, although it contained 24% porosity. The observed high strength for the porous material is very likely due to the presence of pores between grains as well as the grain alignment effects. Toughening of silicon nitride generally arises from crack wake-toughening mechanisms including bridging of the crack by unbroken fibrous grains.[18] In silicon nitride with aligned fibrous grains, fracture resistance is steeply raised in short crack propagation because of effective operation of the grain bridging, leading to the improved strength.[19] It can be hypothesized that the grain bridging is further enhanced by the pores around the aligned fibrous grains, because they facilitate debonding between the grains. In reality, a large number of the pulled-out grains without being broken can be observed, as seen in Figure 2. Additionally, some of the authors recently reported the higher fracture energy of the porous silicon nitride (approximately 500 J/m^2)[1,11] than that of the dense one (approximately 180 J/m^2), where both materials had anisotropic microstructures with aligned grains. The difference in the fracture energy seems to be attributed to the presence of pores. The pores presumably enhance grain pullout and bridging, which leads to higher fracture energy as well as higher strength.[19] Another possible reason for the high strength is the decreased flaw size by sinter forging. Even if the large defects exist in the green compact, their sizes can be substantially reduced by the forging,[15] because the green compact shrinks in the pressing direction. In this work, stress surface of the bending specimen was perpendicular to the pressing direction. Therefore, the grinding scar hardly act as fracture origin. As mentioned above, the porous anisotropic silicon nitride was successfully fabricated by the partial-sinter-forging technique, and the specimen showed relatively high bending strength for the porous material. This technique has applicability as a real fabrication process even for the complicated-shape or large-sized components.

3. Mechanical properties of samples by tape casting of β-Si₃N₄ whisker

Five specimens were used for each characterization, and stable crack growth was obtained until the completion of the test for three specimens of both the PSN and RSN materials, with sufficient reproducibility of the L-D curves. The other two measurements of the RSN material resulted in catastrophic failures during loading. On the other hand, those of the PSN material exhibited large crack deflection along the aligned boundaries from the notch; these occurred when a crack front approached the base of the triangular ligament with the Mode I crack constraint being small. Successful examples of the L-D curves for the two materials are shown in Figure 3. The plot for the RSN material showed a smooth L-D curve, which is normally observed for CNB tests of monolithic ceramic materials, whereas the plot for the PSN material demonstrated serrations throughout almost the entire region. The serrations were similar to the saw-tooth appearance of the L-D curve in the CNB tests of the whisker-reinforced ceramics,[20] which suggests repeated crack initiation and arrest. During loading, any buckling on the compressive half of the specimen was not observed. The effective fracture energy (γ_{eff}) is defined as follows:

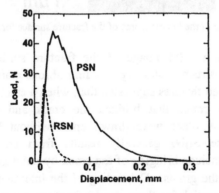

Figure 3 Load-deflection (L-D) diagrams of the chevron-notched bend tests for the PSN and RSN materials.

$$\gamma_{eff} = \frac{W_{WOF}}{2A} \qquad (5)$$

where W_{WOF} is the energy under the L-D curve and A is the area of the specimen. By substituting the W_{WOF} value that is calculated from Figure 3 into Equation 5, one can obtain γ_{eff} values of 490 and 70 J/m^2 for the PSN and RSN materials, respectively. Note that the γ_{eff} value of the PSN material is 7 times larger than that of the RSN material. The former value also is enormously high, in comparison to those obtained for other toughened silicon nitrides, such as 96 J/m^2 for a SiC-whisker-reinforced dense Si$_3$N$_4$[20] and 246 J/m^2 for a highly anisotropic Si$_3$N$_4$ that was produced via superplastic deformation.[21] These values were measured using exactly the same CNB test techniques.

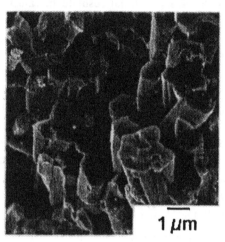

1 μm

Figure 4 SEM micrograph of the ligament area of the fracture surface for the PSN material.

Figure 4 shows a SEM image of the fracture surface of the PSN test specimen. The microstructure shows protruding fibrous grains as well as holes or hollows. The observed features suggested that, when a crack propagated and was opened, the fibrous grains that bridged the crack and had been originally interlocked with each other were drawn apart without being broken. The toughening of dense Si$_3$N$_4$ generally results from crack-wake toughening mechanisms, which include bridging of the crack by unbroken fibrous grains and frictional pullout of the grains. Debonding of the interface between the matrix and the grains is essential for these mechanisms to operate. In dense silicon nitrides where interfaces are relatively strong, however, such debonding is restricted and a limited number of the grains are involved with the toughening mechanisms. On the other hand, in the PSN material, pores around the aligned fibrous grains cause a crack that is propagating perpendicular to the alignment to tilt or twist, which leads to crack deflections.[22] In addition, such crack deflections may facilitate debonding between the interlocking fibrous grains, which leads to several of the drawn-out grains being unbroken, as observed in Figure 4. The large fracture energy obtained in this porous Si_3N_4 is presumably attributable to sliding resistance associated with these drawn-out fibrous grains as well as the crack-deflection effects.

SUMMARY

(1). The porosity content was successfully controlled by the PHP and partial sintering forging method. Mechanical properties, such as Young's modulus, strength, and fracture toughness, decreased with increasing porosity, however, the decrease of strength and toughness is more moderate than for the change in Young's modulus with porosity. The strain tolerance and fracture energy increased with porosity. This slight increase of the fracture energy can be explained by fact that the fracture process is accompanied by the pull-out of the fibrous Si_3N_4 grains and an increase in fracture surface roughness.

(2) The fracture energy of a porous silicon nitride (Si_3N_4) with aligned fibrous grains was investigated using a chevron-notched-beam technique. A crack was constrained to propagate normal to the grain alignment. The obtained fracture energy was 500 J/m^2, which was 7 times larger than that of a dense Si_3N_4 where fibrous grains were randomly oriented. The large fracture energy was attributed to sliding resistance associated with the aligned fibrous grains, as well as crack deflection effects.

ACKNOWLEDGEMENT

This work has been supported by METI, Japan, as part of the Synergy Ceramics Project. Part of the work has been supported by NEDO. The authors are members of the Joint Research Consortium of Synergy Ceramics.

REFERENCES

[1] Y. Shigegaki, M.E.Brito, K.Hirao, M.Toriyama, & S. Kanzaki, "Strain Tolerant Porous Silicon Nitride, *J. Am. Ceram. Soc.*, 80 [2] 495-498 (1997).

[2] J.C Wang,Young's Modulus of Porous Materials, Part 2 Young's Modulus of Porous Alumina with Changing Pore Structure. *J. Mater. Sci.*,19,809-14 (1984)

[3] G.J. Zhang & T. Ohji, "Effect of BN Content on Elastic Modulus and Bending Strength of SiC-BN in situ Ceramics,"*J. Mater.Res.*,15[9]1876-80(2000).

[4] T. Kusunose, Y. H. Choa, T. Sekino, & K. Niihara, "Mechanical Properties of Si_3N_4/BN Composites by Chemical Processing, " *Key Eng. Mater.*, 161-163, 475-479 (1999).

[5]J.B. Davis, A. Kristoffersson, E. Carlström, & W.J. Clegg, "Fabrication and Crack Deflection in Ceramic Laminates with Porous Interlayers," *J. Am. Ceram. Soc.*, 83 [10] 2369-2374 (2000).

[6]A. Díaz, W. Redington & S. Hampshire, "Fabrication of Porous Silicon Nitride and Effects of Controlled Porosity on Mechanical Properties," in *Proceedings of the 5th Symposium on Synergy Ceramics, Tokyo 1-2 January 2001.*

[7]C. Kawai & A. Yamakawa, "Effect of Porosity and Microstructure on the Strength of Si_3N_4: Designed Microstructure for High Strength, High Thermal Shock Resistance, and Facile Machining," *J. Am. Ceram. Soc.*, 80 [10] 2705-2708 (1997).

[8]J.F. Yang, Z.Y. Deng, & T. Ohji, "Fabrication and Mechanical Properties of Porous Si_3N_4 Ceramics Doped with Yb_2O_3 Additions," *J. Eur. Ceram. Soc.*, (in Press)

[9]J.F. Yang, G.J. Zhang & T. Ohji, "Porosity and Microstructure Control of Porous Ceramics by Partial Hot-Pressing," *J. Mater. Res.*, 16 [7] 1916-18 (2001).

[10]N. Kondo, Y. Suzuki & T. Ohji, "High-Strength Porous Silicon Nitride Fabricated by the Sinter-Forging Technique," *J Mater. Res.*, 16 [1] 32-34 (2001).

[11]Y. Inagaki, T. Ohji, S. Kanzaki, & Y. Shigegaki, "Fracture Energy of an Aligned Porous Silicon Nitride," *J. Am. Ceram. Soc.*, 83 [7] 1807-1809 (2000).

[12]W.D. Kingery, H.K. Bowen, & D.R. Uhlmann, *Introduction to Ceramics*, New York, A Wiley-Interscience Publication, pp. 787 (1975)

[13] R. W. Rice, *Porosity of Ceramics*, New York:Marcel Dekker, Inc. (1998)

[14]T. Ostrowski & J. Rödel, "Evaluation of Mechanical Properties of Porous Alumina during Free Sintering and Hot-Pressing," *J. Am. Ceram. Soc.*, 82 [11] 3080-86 (1999).

[15]N. Kondo, E. Sato & F. Wakai, "Geometrical Microstructural Development in Superplastic Silicon Nitride with Rod-Shaped Grains," *J. Am. Ceram. Soc.* 81 [12] 3221-3227 (1998).

[16]N. Kondo, Y. Suzuki, & T. Ohji, "Superplastic Sinter-Forging of Silicon Nitride with Anisotropic Microstructure Formation," *J. Am. Ceram. Soc.*, 82 [4] 1067-1069 (1999).

[17]J.C. Hay, E.Y. Sun, G.M. Pharr, P.F. Becher, & K.B. Alexander, "Elastic Anisotropy of Beta-Silicon Nitride Whiskers," *J. Am. Ceram. Soc.*, 81 [10] 2661-2669 (1998).

[18]P.F. Becher, "Microstructural Design of Toughness Ceramics," *J. Am. Ceram. Soc.*, 74 [2] 255-269 (1991).

[19]T. Ohji, K. Hirao, & S. Kanzaki, "Fracture-Resistance Behavior of Highly Anisotropic Silicon-Nitride," *J. Am. Ceram. Soc.*, 78 [11] 3125-3128 (1995).

[20]T. Ohji, Y. Goto, & A. Tsuge "High-Temperature Toughness and Tensile Strength of Whisker-Reinforced Silicon Nitride," *J. Am. Ceram. Soc.*, 74 [4] 739-745. (1991).

[21] N. Kondo, Y. Inagaki, Y. Suzuki, and T. Ohji, "High-Temperature Fracture Energy of Superplastically Forged Silicon Nitride Pulsed Electric," *J. Am. Ceram. Soc.*, 84 [8] 1791-1796 (2001).

[22]K.T. Faber & A.G. Evans, "Crack Deflection Processes I. Theory. *Acta Metall.*," 31 [4] 565-576 (1983).

LIQUID PHASE SINTERING OF SiC WITH AlN AND RARE-EARTH OXIDE ADDITIVES

M. Balog, P. Šajgalík and Z. Lenčéš
Institute of Inorganic Chemistry,
Slovak Academy of Sciences
Dubravska cesta 9
842 36 Bratislava, Slovakia

J. Kečkéš
Erich Schmid Institute for Materials Research, Austrian Acad. Sciences,
Jahnstrasse 12, A-8700 Leoben Austria

J.-L. Huang
Department of Materials Science and Engineering
National Cheng Kung University
Tainan, Taiwan 701
Republic of China

ABSTRACT

Liquid phase sintering of SiC was performed using rare-earth oxide R_2O_3 (R = Y, Yb, and Sm) and AlN sintering additives. The addition of 2 vol% β-SiC seeds or α-SiC platelets resulted in bimodal microstructure after annealing for 10 hours at 1850°C in mixed N_2 + Ar atmosphere. Samples had a negligible weight loss after hot pressing. The bimodal microstructure contributed to the improvement of fracture toughness. Seeding by α-SiC seeds stimulated the $\beta \rightarrow \alpha$-SiC phase transformation and increased the hardness. Samples with α-SiC seeds and complex Yb_2O_3 + Sm_2O_3 + AlN additives had a Vickers hardness of 27.5 GPa. Seeding by β-SiC whiskers hindered the $\beta \rightarrow \alpha$-SiC phase transformation and increased the fracture toughness up to 6.5 MPa·m$^{1/2}$.

INTRODUCTION

Silicon carbide (SiC) is a promising material for high temperature application due to its high strength, good wear and corrosion resistance at elevated temperatures. SiC can be densified via solid-sate sintering with the addition of B and C at temperatures around 2100°C.[1] However, in comparison with liquid phase sintered Si_3N_4 based ceramics, the solid state sintered SiC has a lack fracture toughness for engineering applications (K_{IC} < 3 MPa·m$^{1/2}$). In the past two decades the traditional additives like boron, carbon and aluminium, used for solid state sintering of SiC were replaced by metal oxides.[2-5] This allowed the liquid phase sintering of SiC at relatively lower temperatures (1850°C to 2000°C). These materials exhibit higher strength and fracture toughness compared to solid state sintered SiC.[5] The fracture toughness of liquid phase sintered SiC with Y_2O_3 and Al_2O_3 additives reached the value of 7 MPa·m$^{1/2}$.[6,7] The sintered bodies had a microstructure with elongated α-SiC platelets after annealing at high temperatures ~1900°C for several hours. On the other hand, the densification of SiC with oxide additives is accompanied by high weight losses up to 15 wt% depending on the total additive content, including the SiO_2 impurities in the starting SiC powder.[8] The results showed that the main contributors to weight loss are SiO_2 and Al_2O_3, due to the reactions with SiC (or with free carbon, present in SiC starting powder as an impurity) producing gaseous species like SiO and Al_2O, respectively. Moreover, the gaseous reaction products can hinder the densification. Although SiO_2 is always present at the surface of SiC starting powder, Al_2O_3 can be replaced with AlN. Recently, Hoffmann et al.[8] and Keppeler et al.[9] succeeded to densify SiC with Y_2O_3 and AlN sintering additives. Samples had a negligible weight loss.

The binary system AlN – Y_2O_3 has a eutectic temperature of 1730°C under 1 MPa of N_2 pressure.[10] Some of the AlN – R_2O_3 systems showed similar phase diagram, which contains a liquid region with a steep liquidus line from R_2O_3 and a gaseous region on the AlN side.[11] Jackson et al.[12] used lanthanide oxides along with Y_2O_3, for the densification of AlN. The lanthanide oxides were added in equimolar amounts to Al_2O_3 impurity in AlN powder, and were effective densification aids for AlN when sintered at 1850°C.

In the system SiC – AlN, a solid solution with up to 50% SiC was formed by heating the appropriate starting mixtures up to 2000°C.[13] However, SiC and AlN could form solid solution over 1800°C.[14] Rafaniello et al. prepared dense SiC – AlN samples by hot pressing at 2100°C. Samples had a bimodal microstructure, which consist of SiC-rich small grains and AlN-rich large grains.[15] By heating these samples up to 2300°C, homogeneous SiC–AlN solid solution was obtained. On the other hand, it has been shown that a complete solid solution between SiC, AlN, and Al_2OC can be formed by heating an intimate mixture of reactants already at 1600°C.[16]

Different rare-earth oxides (Y_2O_3, Yb_2O_3, Sm_2O_3) and their combinations with AlN are used as sintering additives for the densification of SiC in this study. The present work has investigated the effectiveness of this additive system on the liquid phase sintering of SiC and on the microstructural development. From this reason the starting powders were also doped with α- or β-SiC seeds to promote grain growth. Hardness and fracture toughness of sintered materials are evaluated in more details.

EXPERIMENTAL

Fine-grain β-SiC powder (Superior Graphite, USA) was mixed with rare-earth oxides Y_2O_3 (grade C, H.C. Starck, Germany), Yb_2O_3 (H.C. Starck, Germany), Sm_2O_3 (Russia), and AlN (grade H, Tokuyama Co., Japan) in a ratio listed in Table I. Further, α-SiC (grade SF, C-Axis Technology, Canada) or β-SiC (grade T-1, grain size 2.5 µm, aspect ratio 6, Tokai Carbon, Japan) seeds were added to the powder mix.

Table I. Chemical composition of samples

Sample	Composition (wt. %)					
	SiC	SiC seeds	AlN	Y_2O_3	Yb_2O_3	Sm_2O_3
SCα-Y	85	2 (α)	3	10	-	-
SCα-YYb	85	2 (α)	3	4.59	5.41	-
SCα- YSm	85	2 (α)	3	3.93	-	6.07
SCα-YbSm	85	2 (α)	3	-	5.31	4.69
SCβ- Y	85	2 (β)	3	10	-	-
SCβ- YYb	85	2 (β)	3	4.59	5.41	-
SCβ- YSm	85	2 (β)	3	3.93	-	6.07
SCβ- YbSm	85	2 (β)	3	-	5.31	4.69

The powder mixtures were ball milled in isopropanol with SiC balls for 24 hours. The homogenized suspension was dried, sieved through 25 μm sieve screen and hot pressed at 1850°C for 1h under pressure of 30 MPa in Ar+N_2 atmosphere. The hot-pressed samples were further annealed at 1850°C for 10 hours to enhance grain growth.

Bulk densities were measured by Archimedes method in mercury. The crystalline phases were determined by X-ray diffraction (XRD) on the ground samples. The microstructures were observed by scanning electron microscopy (SEM). For this purpose the sintered samples were cut, polished, and plasma etched with CF_4 + O_2 gas mixture. Vickers hardness and fracture toughness were measured using LECO Hardness tester at 9.8 N and 98 N loads, respectively.

RESULTS AND DISCUSSION
Influence of additives and seeds on the microstructure
All specimens reached 95-99% of theoretical density after hot pressing at 1850°C for 1 hour. Generally, samples with Sm_2O_3 addition had lower final density compared to other ones. Relatively high densities of sintered SiC samples indicate that the investigated additives are suitable for liquid phase sintering of SiC, however, the temperature schedule of sintering should be further optimized.

The microstructure of sintered SiC samples is shown in Figs. 1 and 2. Generally, all samples exhibits homogeneous microstructure with equiaxed grains after 1 hour sintering. The core-rim structure of SiC grains was frequently observed in all samples. It means that SiC grains grow by the solution-reprecipitation mechanism. The kind of rare-earth oxide (R_2O_3) additive does not have strong influence on the microstructure. On the other hand, comparison of samples with the same additives, but with α- or β-SiC seeds shows that samples with β-SiC seeds have a thicker rim and in average larger grains compared to α-SiC doped samples. It reflects different growth behavior of SiC grains. Kim et al.[17] showed that β-SiC has higher solubility than α-SiC in Y_2O_3 – Al_2O_3 – SiO_2 based liquid phase. Although, in present work the additive system is composed of R_2O_3 – AlN – (SiO_2), certain similarity with higher solubility of β-SiC in transient liquid might be expected.

Table II. Density (ρ, ρ_0) and phase composition of SiC samples after annealing at 1850°C for 10 hours

Sample	$\rho - \rho_0$ (%)	$\alpha/(\alpha+\beta)$ SiC ratio (%)
SCα-Y	-0.9	64
SCα-YYb	-4.7	32
SCα- YSm	-5.0	30
SCα-YbSm	-5.4	11
SCβ- Y	-2.1	15
SCβ- YYb	-1.8	7
SCβ- YSm	-3.0	13
SCβ- YbSm	-2.2	6

Samples were annealed after hot pressing for 10 hours at 1850°C in mixed N_2+Ar atmosphere. Substantial grain growth was not observed for the samples doped with α-SiC seeds, as it is shown in Figs. 1a and 3a. On the other hand, the average grain size of β-doped samples increased from 0.7 μm to 2-7 μm after

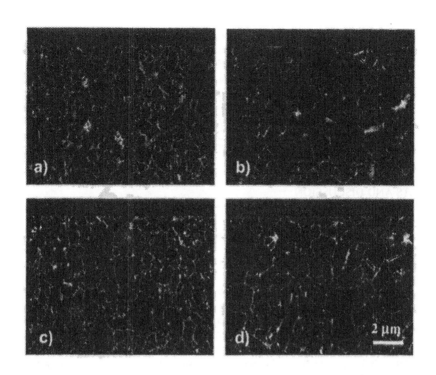

Fig. 1: Microstructure of SiC samples after hot-pressing at 1850 °C for 1h:
a) SCα-Y, b) SCβ-Y, c) SCα-YYb, b) SCβ-YYb.

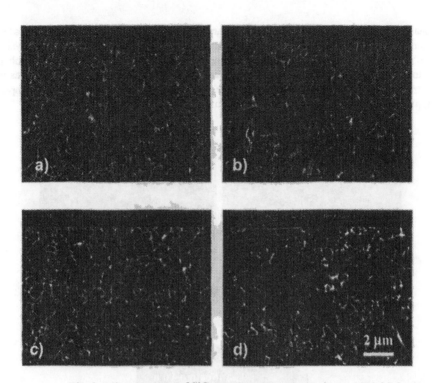

Fig. 2: Microstructure of SiC samples after hot-pressing at 1850 °C for 1h:
a) SCα-YSm, b) SCβ-YSm, c) SCα-YbSm, b) SCβ-YbSm.

annealing for 10 hours, certify Figs. 2b and 3b. The annealing of SiC samples was accompanied by decrease of density in the range 0.9-5.5% (Table II). The EDS analysis of SCα-Y sample showed the transport of Y^{3+} ions towards the surface (Fig.4), most probably due to the concentration/temperature gradient across the sample. The sample surface was free of yttrium up to the depth of 150 μm. However, the weight loss is remarkably lower compared to liquid-phase sintered SiC with Al_2O_3 containing additives.

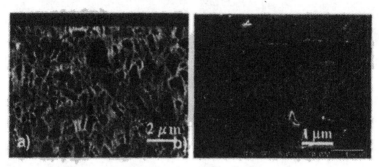

Fig. 3. Microstructure of SiC samples after annealing at 1850°C for 10 hours: a) SCα-Y, b) SCβ-YSm.

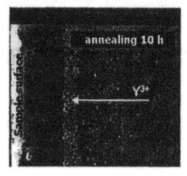

Fig. 4. EDS analysis of SCα-Y sample showing the migration of Y^{3+} towards the surface after annealing at 1850°C for 10 hours.

Influence of additives and seeds on the β → α-SiC phase transformation

The $\alpha/(\alpha+\beta)$-SiC phase ratio was calculated by the method described by Ruska et al.[18] and by Tanaka et al.[19] According the results listed in Table II the used additives have a weak influence on β → α-SiC phase transformation. Concerning the additives, the α-phase content was higher in Y_2O_3 containing samples compared to materials with Yb_2O_3 addition. Most probably it is due to the higher refractoriness of ytterbia-based glassy phase, because the eutectic temperature of Y_2O_3 - $Y_2O_3 \cdot SiO_2$ system is 1800°C, while for the Yb_2O_3 - $Yb_2O_3 \cdot SiO_2$ it is 1850°C. Samples were annealed at 1850°C in this work, and the higher viscosity of Yb_2O_3–AlN-Al_2O_3-SiO_2 based transient liquid phase hindered the β → α-phase transformation.

Table II shows that the kind of seed particles has a stronger influence on β → α-SiC phase transformation compared to different R_2O_3 additives. Comparison of samples with the same Y_2O_3-AlN sintering additives showed that the final α-SiC content of the samples with 2 wt% α-seeds was about 40%, while with β-seeds only 12% after annealing at 1850°C for 10 hours. These results are in agreement with the observation of Kim et al.[20] that the polytype distribution in SiC materials is related preferentially to the polytype distribution in starting powders, rather than the chemistry of sintering aids. It can be concluded that it is easier to control the final phase composition of SiC by deliberately adding a small amount of seeds (1–3 wt%) to the starting powders. After annealing at high temperatures (T ≥ 1850°C) self-reinforced microstructure consisting of large platelet-like elongated grains and relatively small matrix grains can be obtained.

Finally, it should be pointed out that the samples were intentionally annealed in mixed N_2 + Ar atmosphere. The main reason for the selection of mixed atmosphere was to keep the partial pressure of N_2 on the required level and avoid the decomposition of AlN. On the other hand, nitrogen partly influences the microstructural development. It has been published by several authors, [21,22] and also our previous results confirmed that nitrogen hinders the β → α-SiC phase transformation and grain growth.[23] The influence of N_2 on the microstructural development was not the subject of this work.

Mechanical properties

The Vickers hardness (HV1) and indentation fracture toughness (K_{IC}) of SiC ceramics are shown in Figs. 5 and 6. The results show that the hardness is dependent on the kind of added α- or β-SiC seeds. Samples doped with α-SiC seeds exhibit higher hardness compared to β-doped ceramic materials. It was expected that the samples with higher α-SiC content would have a little bit higher hardness, because the Knoop hardness of α-SiC is 29 GPa ((0001) face of 6H

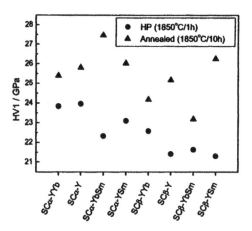

Fig. 5. Vickers hardness of hot pressed and annealed samples at 1850°C.

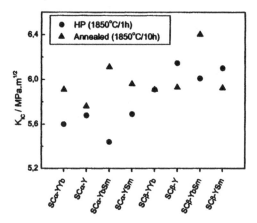

Fig. 6. Fracture toughness of hot pressed and annealed samples at 1850°C.

polytype), and β-SiC is 28 GPa ((100) face of 3C structure).[24] Samples with α-seeds have also finer microstructure. On the other hand, samples doped with β-SiC seeds have a coarser mictrostructure and higher fracture toughness. The elongated shape of SiC grains grown on β-seeds has a beneficial effect on the fracture toughness of sintered samples.

The hardness was measured also by nanoindentation, because by micro-hardness measurement several tens of grains and also grain boundary phase are tested and provides an average value from the tested area. As it was mentioned above, all samples had a core-rim structure and an example is shown in Fig. 7a. On the base of SEM observations it is expected that the rim is the solid solution of SiC-AlN-(R_2O_3). Although in present work only 3 wt% AlN was added to SiC and the firing temperature was relatively low (1850°C) for solid solution of SiC and AlN, more detailed study of the core and rim of individual grains was carried out

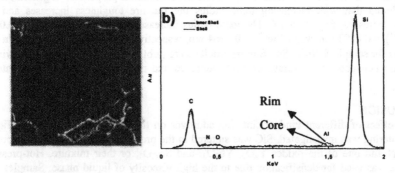

Fig. 7. SEM micrograph of the core-rim structure (a) and the EDS analysis (b)   of SCα-Y sample.

by nanoindentation. The average values of nanohardness calculated from 60 measurements on sample SCα-YSm were 39.3 ± 2.4 GPa and 27.4 ± 3.2 GPa for the core and rim, respectively. The difference is remarkable and indicates to the different composition of the core and rim of SiC grains. The EDX analysis of the core and rim of SCα-Y sample is shown in Fig. 7b. The rim contains a small amount of Al, while the core is free of aluminum. Part of AlN was incorporated into the rim during the dissolution-reprecipitation process. The decrease of hardness with the presence of Al in SiC grains is in agreement with the results of Ruh et al.[25] These authors described that the hardness values decreased linearly in the SiC-AlN solid solution region with increasing AlN content.

Fig. 8 shows the comparison of nanohardness and microhardness data. Each value of nano- and micro-hardness is the average value of 60 and 20 indentation for each sample. Nanohardness does not follow the trends observed for microhardness.

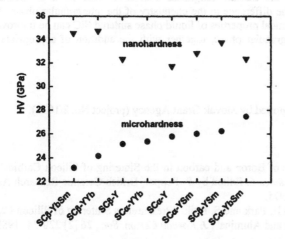

Fig. 8. Nanohardness and microhardness annealed samples.

It is generally accepted that the mechanical properties depend also on the chemistry of grain boundary phase. The analysis of measured data for HV and K_{IC} shows a correlation between the ionic radius of rare-earth dopant and the mechanical properties. In general, it can be concluded that with increasing ionic radius the fracture toughness increases and the hardness decreases, Figs. 5 and 6. The ionic radius increases in the order: $r(Yb^{3+}) = 0.0858$ nm, $r(Y^{3+}) = 0.0893$ nm, and $r(Sm^{3+}) = 0.0964$ nm, respectively.[26] The only exception in this trend was for sample SCα-YbSm. Similar results were published by Zhou et al.,[27] showing a correlation between atomic radius of R_2O_3 additives and mechanical properties of sintered SiC.

CONCLUSION

The effect of different kind of sintering additives on the densification of SiC by liquid phase sintering was investigated. SiC was sintered in the presence of the liquid phase formed from AlN and one of the oxides Y_2O_3, Yb_2O_3, and Sm_2O_3, or their mixture. Hot-pressing technique was used for densification due to the high viscosity of liquid phase. Samples had lower weight loss, because the AlN additive was stabilized by N_2 containing atmosphere. The density of samples was in the range of 94-99% theoretical density, depending on the kind of sintering additives. All the investigated R_2O_3 – AlN sintering additive systems are suitable for densification of SiC at temperatures < 1900°C.

The SEM analyses of hot pressed samples showed a core-rim microstructure. Most probably, it is due to the partial dissolution and subsequent precipitation of SiC in the transient liquid phase.

The microstructure and mechanical properties are dependent also on the kind of used α- or β-SiC seeds. Seeding by α-SiC seeds stimulated the β → α-SiC phase transformation and increased the hardness. Samples with α-SiC seeds and Yb_2O_3 + Sm_2O_3 + AlN additives had a Vickers hardness of 27.5 GPa and indentation fracture toughness 6.1 MPa·m$^{1/2}$. Seeding by β-SiC whiskers hindered the β → α-SiC phase transformation and increased the fracture toughness. The highest fracture toughness of 6.5 MPa·m$^{1/2}$ had the sample doped with β-seeds and with Yb_2O_3 + Sm_2O_3 + AlN additives annealed for 5 hours at 1850°C.

It was observed that with decreasing ionic radius of rare-earth additive the hardness of SiC ceramics increased and the fracture toughness decreased. This trend of mechanical properties was attributed to the difference in the chemistry of the intergranular phase. These results suggest that the mechanical properties of liquid phase sintered SiC can be improved by the optimization of the intergranular phase, together with the addition of appropriate seed particles.

ACKNOWLEDGMENTS

The research is partly supported by Slovak Grant Agency (project No. 2/1033).

REFERENCES

1) S. Prochazka, "The role of Boron and carbon in the Sintering of Silicon Carbide", pp. 171-181 in *Special Ceramics 6*. Edited by P. Popper, British Ceramic Research Assoc., Stoke-on-Trent, U.K., 1975.

2) D.H. Kim, C.W. Jang, B.H. Park and S. Baik, "Pressureless Sintering of Silicon Carbide with Additions of Yttria and Alumina", *J. Korean Ceram. Soc.*, **26** [2] 228-34 (1989).

3) A.K. Misra, "Thermochemical Analysis of the Silicon Carbide – Alumina Reaction with Reference to Liquid-Phase Sintering of Silicon Carbide", *J. Am. Ceram. Soc.*, **74** [2] 345-51 (1991).

4) M.A. Mulla and V.D. Krstic, "Low-Temperature Pressureless Sintering of β-Silicon Carbide with Aluminum Oxide and Yttrium Oxide Additions", *Am. Ceram. Soc. Bull.*, **70** [3] 439-43 (1991).

5) N.P. Padture, "In Situ-Toughened Silicon Carbide," *J. Am. Ceram. Soc.*, **77** [2] 519-23 (1994).

6) D.H. Cho, Y.W. Kim and W.J. Kim, "Strength and Fracture Toughness of In Situ-Toughened Silicon Carbide", *J. Mater. Sci.*, **32**, 4777-4782 (1997).

7) N.P. Padture and B.R. Lawn, "Toughness Properties of a Silicon Carbide with an in Situ Induced Heterogeneous Grain Structure" *J. Am. Ceram. Soc.*, **77** [10] 2518-22 (1994).

8) M.J. Hoffmann and M. Nader, "In Situ Toughening of Non Oxide Ceramics – Opprtunities and Limits," pp. 133-146 in *Engineering Ceramics '96*. Edited by. G.N. Babini, M. Haviar, P. Šajgalík. Kluwer Academic Publ., Netherlands, 1997.

9) M. Keppeler, H.G. Reichert, J.M. Broadley, G. Thurn and I. Wiedmann, "High Temperature Mechanical Behaviour of Liquid Phase Sintered Silicon Carbide," *J. Europ. Ceram. Soc.*, **18**, 521-26 (1998).

10) Z.K. Huang and T.Y Tien, "Solid-Liquid Reaction in the Si_3N_4-AlN-Y_2O_3 System under 1 MPa of Nitrogen," *J. Am. Ceram. Soc.*, **79** [6] 1717-19 (1996)

11) Z.K. Huang, D.S. Yan, and T-Y Tien, "Compound Formation and Melting Behavior in the AB Compound and Rare Earth Oxide Systems, " *J. Solid State Chem.*, **85**, 51 (1990).

12) T.B. Jackson, A.V. Virkar, K.L. More, R.B. Dinwiddie Jr. and R.A. Cutler, "High-Thermal-Conductivity Aluminum Nitride Ceramics: The Effect of Thermodynamic, Kinetic, and Microstructural Factors," *J. Am. Ceram. Soc.*, **80** [6] 1421-35 (1997).

13) V.L. Matkovich, E. Colton and J.L. Peret, "Refractory Materials and the Method of Making Same, " US Pat. No. 3 259 509, July 5, 1966.

14) I.B. Cutler and P.D. Miller, "Solid Solution and Process for Producing a Solid Solution," US Pat. No. 4141740, Febr. 27, 1979.

15) W. Rafaniello, M.R. Plichta and A.V. Virkar, "Investigation of Phase Stability in the System SiC-AlN," *J. Am. Ceram. Soc.*, **66** [4] 272-276 (1983).

16) I.B. Cutler, P.D. Miller, W. Rafaniello, H.K. Park, D.P. Thompson and K.H. Jack, "New Materials in the Si-C-Al-O-N and Related Systems," *Nature*, **275** [5679] 434-35 (1978).

17) Y.W. Kim, M. Mitomo, H. Emoto and J.G. Lee, "Effect of Initial α-Phase Content on Microstructure and Mechanical Properties of Sintered Silicon Carbide," *J. Am. Ceram. Soc.*, **81** [12] 3136-40 (1998).

18) J.Ruska, L.J. Gauckler, J. Lorentz and H.U. Rexer, "The Quantitative Calculation of SiC Polytypes from Measurements of X-ray Diffraction Peak Intensities," *J. Mater. Sci.*, **14**, 2013-17 (1979).

19) H. Tanaka and N. Iyi, "Simple Calculation of SiC Polytype Contents from Powder X-Ray Diffraction Peaks," *J. Ceram. Soc. Japan*, **101** [11] 1313-14 (1993).

20) Y.W. Kim, M. Mitomo and H. Hirotsuru, "Microstructure and Polytype of In Situ-Toughened Silicon Carbide," *Korean Journal of Ceramics*, **2** [3] 152-156 (1996).

21) Y.W. Kim and M. Mitomo, "Fine-Grained Silicon Carbide Ceramics with Oxynitride Glass," *J. Am. Ceram. Soc.*, **82** [10] 2731-36 (1999).

22) H.W. Jun, H.W. Lee, G.H. Kim, H.S. Song and B.H. Kim, "Effect of Sintering Atmosphere on the Microstructure Evolution and Mechanical Properties of Silicon Carbide Ceramics", *Ceram. Eng. Sci. Proc.*, **18** [4] 487-505 (1997).

23) M. Balog, Z. Lenčéš and P. Šajgalík, "Liquid Phase Sintering of SiC," pp. 53-57 in Proc. *4ᵗʰ Conf. Processing of Ceramic Materials*. Edited by B. Plešingerová and T. Kuffa. Hutnícka fakulta TU, Košice, Slovakia, 2001.

24) P.T.B. Shaffer, "Engineering Properties of Carbides", pp. 804-811 in *Ceramics and Glasses*, Engineered Materials Handbook, Vol. 4. Edited by S.J. Schneider Jr., ASM International, USA, 1991.

25) R. Ruh and A. Zangvil, "Composition and Properties of Hot-Pressed SiC-AlN Solid Solutions," *J. Am. Ceram. Soc.*, **65** [5] 260-65 (1982).

26) F.H. Spedding and K. Gschneidner, "Crystal Ionic Radii of the Elements," pp. F-152 in Handbook of Chemistry and Physics, 51ˢᵗ ed. Edited by R.C. Weast, The Chemical Rubber Co., Cleveland, Ohio, 1971.

27) Y. Zhou, K. Hirao, M. Toriyama, Y. Yamauchi and S. Kanzaki, "Effects of Intergranular Phase Chemistry on the Microstructure and Mechanical Properties of Silicon Carbide Ceramics Densified with Rare-Earth Oxide and Alumina Additions," *J. Am. Ceram. Soc.*, **84** [7] 1642-44 (2001).

Silicon-Based Structural Ceramics

EFFECT OF ADDITIVES ON MICROSTRUCTURAL DEVELOPMENT AND MECHANICAL PROPERTIES OF LIQUID-PHASE-SINTERED SILICON CARBIDE DURING ANNEALING

You Zhou, Kiyoshi Hirao, Yukihiko Yamauchi and Shuzo Kanzaki
Synergy Materials Research Center, National Institute of Advanced Industrial Science and Technology (AIST), Nagoya 463-8687, Japan

ABSTRACT

Four dense SiC ceramics sintered with additions of various rare-earth oxides (RE_2O_3, RE = La, Nd, Y and Yb) in conjunction with alumina were annealed at high temperatures. The SiC ceramics exhibited quite different behaviors of microstructural development during annealing which strongly affected the mechanical properties of the annealed materials. While choosing La_2O_3 or Nd_2O_3, the grain growth during annealing was moderate and highly anisotropic, which contributed to simultaneous improvement of flexural strength and fracture toughness; in contrast, the addition of Y_2O_3 or Yb_2O_3 caused too large grain coarsening which resulted in degraded mechanical strength.

INTRODUCTION

Silicon carbide can be densified either by a solid-state mechanism, where boron-carbon system sintering additives are used,[1,2] or by a liquid-phase sintering mechanism, where oxide[3-5] or non-oxide[6,7] sintering additives are employed. Nowadays, the solid-state-sintering of SiC containing B-C additives has become a key technology in manufacturing SiC dense bodies for applications as mechanical components, however, there still are some problems associated with this process, e.g., the low fracture toughness due to the strong grain-bonding which impedes toughening mechanisms such as crack deflection and grain bridging, and the frequent occurrence of exaggerated grain growth owing to the very high sintering temperatures which are over 2100 °C. Recently, it has been reported that those shortcomings of the solid-state-sintered SiC might be overcome by the approach of liquid phase sintering.[4-10] The liquid-phase-sintered SiC may form an *in-situ* toughened microstructure, consisting of elongated SiC grains surrounded by weakly-bonded intergranular phases, which promotes crack-wake bridging and then results in an increase in the toughness.

As for the oxide sintering additives for LPS-SiC, mixtures of Al_2O_3 and Y_2O_3 have predominantly been used so far. In literature, work on sintering of SiC using other oxides as additives is very limited. Recently,[11] we found that a variety of rare-earth oxides (RE_2O_3) were as effective as the popularly used Y_2O_3 in aiding densification of SiC, and the SiC ceramics containing different RE_2O_3 additives had different mechanical properties although their microstructures were very similar. The changes of the mechanical properties revealed that a decrease in the cationic radius of the RE_2O_3 (RE = La, Nd, Y and Yb) was accompanied by an increase in Young's modulus, hardness and flexural strength and a decrease in fracture toughness of the SiC ceramics. That trend was attributed to the difference in the chemistry of the intergranular phases. On the other hand, it is noticed that those materials all had very fine and equiaxed microstructures owing to the low densification temperature. The fine and equiaxed grains could not effectively contribute to toughening effect such as grain bridging.[12] Therefore, in this work, those dense SiC were annealed at higher temperatures to promote grain growth, with the expectation of obtaining *in-situ*-toughened SiC materials consisting of large elongated grains. The effects of the various RE_2O_3 additives on the microstructural development during annealing and on the mechanical properties of the annealed materials were studied.

EXPERIMENTAL PROCEDURE

The starting powder was high purity β–SiC (UF, Ibiden Co., Gifu, Japan) with an average particle size of 0.30 μm and specific surface area of 20.0 m^2/g. The main impurity was 0.25 wt% oxygen. The additives were Al_2O_3 (AKP50, Sumitomo Chemicals, Tokyo, Japan) and four rare-earth oxides: La_2O_3, Nd_2O_3, Y_2O_3 and Yb_2O_3 (99.9% pure, Nippon Yttrium Co., Tokyo, Japan). Four batches of powders (see Table I) were prepared, each containing 95 vol% SiC and 5 vol% (Al_2O_3 + RE_2O_3) (RE = La, Nd, Y and Yb). The additive composition (molar ratio of Al_2O_3 to RE_2O_3) of each batch was selected at the lowest eutectic point in the Al_2O_3-RE_2O_3 phase diagram. The specimen designations are given in Table I.

The mixtures of SiC and additives were blended in methanol in an SiC planetary mill, then dried with a rotary evaporator, crushed, and screened through a 60-mesh sieve. The mixed powder was hot-pressed at 1800 °C for 1 h under a pressure of 40 MPa in Ar. The final hot-pressed block size was 42 mm x 47 mm x 5 mm. For each composition, more than six blocks were prepared. Some of the hot-pressed blocks were annealed at 1850 or 1950 °C for 3 h under an atmosphere of argon in a graphite resistance furnace.

Bulk densities were measured by the Archimedes method. The hot-pressed and the annealed specimens were cut, polished, and etched by a plasma of CF_4 containing 10% O_2 in a commercial plasma etching apparatus (Model PR31Z, Yamato Scientific, Tokyo, Japan). The microstructures of the etched surfaces were

Silicon-Based Structural Ceramics

observed by a scanning electron microscope (SEM) (Model JSM-6340F, JEOL, Tokyo, Japan) equipped with a field emission gun. Quantitative analyses on digitized SEM photographs were carried out using an image analysis software (Scion Image, Scion Corporation, Frederick, USA). For each specimen at least 2000 grains were measured. The area (A_{grain}), shortest and longest diagonals of each grain in a sectioned plane were automatically determined by the software. The equivalent grain diameter (D) of each grain was calculated by the equation D = $2(A_{grain}/\pi)^{1/2}$, [13] and it was defined as the grain size in this study. The grain-size distribution was evaluated by plotting the fractional cumulative area of the grains versus the corresponding calculated equivalent grain diameters. The mean grain size was considered to be the value for one-half the cumulative area.[14] Considering an elongated grain morphology, the thickness (T) and length (L) of each grain were obtained from the shortest and the longest diagonals in a two-dimensional image, respectively. The mean value of the 10% highest apparent aspect ratios (L/T) was considered to be the mean of the actual aspect ratios (R_{95}).[9, 14, 15]

Table I Compositions and Specimen Designations of Various Mixtures

Specimen designation	RE$_2$O$_3$	Ionic radius of RE^{3+} (nm)	Composition (vol%)		
			β–SiC	Al$_2$O$_3$	RE$_2$O$_3$
AlLa5	La$_2$O$_3$	0.1061	95	3.19	1.81
AlNd5	Nd$_2$O$_3$	0.0995	95	3.25	1.75
AlY5	Y$_2$O$_3$	0.0892	95	3.28	1.72
AlYb5	Yb$_2$O$_3$	0.0858	95	3.49	1.51

X-ray diffraction analysis (XRD) (RINT2500, Rigaku, Tokyo, Japan) was performed on ground powders of the ceramics for phase identification. Test beams with dimensions of 4 mm x 3 mm x 42 mm were sectioned and ground with a 400-grit diamond grinding wheel. The beams were tested in a four-point bending jig with an outer span of 30 mm, an inner span of 10 mm, and at a crosshead speed of 0.5 mm/min. The tensile surfaces of the specimens were polished to a 1/2 µm diamond finish, and the edges were chamfered to reduce edge flaws. Fracture toughness and hardness were measured by the indentation fracture (IF) method using a Vickers indenter at 98 N load and calculated using the equations proposed by Anstis et al..[16] Young's modulus was measured by the pulse-echo method. The indented surfaces were also plasma-etched and observed using SEM in order to examine the crack / microstructure interactions.

RESULTS

Microstructural Development

All specimens were nearly fully-densified. Figure 1 shows the microstructures of the four hot-pressed SiC (AlLa5, AlNd5, AlY5 and AlYb5) and the materials annealed at 1850 or 1950 °C for 3 h. All the four hot-pressed materials had fine and equiaxed microstructures (Figs. 1 (a) to (d)).

After being annealed at 1850 °C, grain growth occurred for all materials (Figs. 1(e) to (h)). However, the extent of grain coarsening and the resulted grain morphologies between materials were different. In the annealed AlLa5 and AlNd5, there were a few large elongated grains. In the annealed AlYb5, almost all grains were large and elongated. In contrast, almost all grains in the annealed AlY5 remained to be equiaxed except that very few elongated grains occasionally appeared in the microstructure.

After being annealed at 1950 °C, grain sizes became even larger (Figs. 1(i) to (l)). In the 1950°C-annealed AlLa5 and AlNd5, nearly all grains have grown to be elongated. In the 1950° -annealed AlYb5, the elongated grains were thicker than those in the 1850°C-annealed AlYb5. The 1950°C-annealed AlY5 consisting of very large plate-like grains, turned to have the coarsest microstructure among the four 1950°C-annealed materials, although the microstructure of the 1850°C-annealed AlY5 was the finest among the four 1850°C-annealed materials.

Mechanical Properties

The flexural strength and fracture toughness of the hot-pressed, 1850°C- and 1950°C-annealed SiC ceramics are listed in Table II. For all the materials, an annealing treatment always had a positive impact on the fracture toughness, and, the higher the annealing temperature, the higher the toughness. In contrast, the influence of annealing on the flexural strength might be positive or negative, depending on material and annealing temperature. After being annealed at 1850°C, the flexural strength of all materials improved. With the annealing temperature increasing to 1950°C, however, while the strength of the annealed AlLa5 and AlNd5 further improved, the annealed AlY5 and AlYb5 suffered a great strength-degradation.

Among all the materials, the 1950°C-annealed AlLa5 and AlNd5 had the best mechanical properties. The flexural strength and fracture toughness of the 1950°C-annealed AlLa5 were 661 MPa and 5.5 MPa $\cdot$ m$^{1/2}$, respectively; and the values for the 1950°C-annealed AlNd5 attained 620 MPa and 5.5 MPa $\cdot$ m$^{1/2}$, respectively.

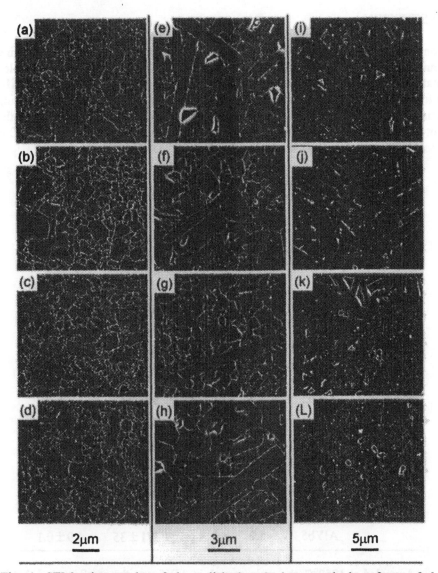

2μm 3μm 5μm

Fig. 1. SEM micrographs of the polished and plasma-etched surfaces of the hot-pressed SiC: (a) AlLa5, (b) AlNd5, (c) AlY5 and (d) AlYb5; the 1850°C-annealed SiC: (e) AlLa5, (f) AlNd5, (g) AlY5 and (h) AlYb5; and the 1950°C-annealed SiC: (i) AlLa5, (j) AlNd5, (k) AlY5 and (L) AlYb5.

DISCUSSION

As mentioned above, the SiC ceramics doped with different RE_2O_3 in combination with Al_2O_3 exhibited different grain growth behaviors during annealing, which in turn affected the mechanical properties of the annealed materials. An image analysis on the micrographs in Fig.1 yielded more specific information on how the microstructures of the various materials developed during annealing. Table II lists the mean grain sizes and mean aspect ratios of the hot-pressed and the annealed materials. The aspect ratio of the hot-pressed SiC is considered to be 1 for equiaxed microstructure.

Table II Mean Grain size, Mean Aspect Ratio, Flexural Strength and Fracture Toughness of the Hot-Pressed and the Annealed SiC Ceramics

Materials		Mean grain size (μm)	Mean aspect ratio	Flexural Strength (MPa)	Fracture toughness (MPa · m$^{1/2}$)
Hot-pressed SiC	AlLa5	0.80	1	504±48	4.2±0.2
	AlNd5	0.91	1	550±42	3.9±0.1
	AlY5	0.76	1	587±35	3.7±0.2
	AlYb5	0.87	1	652±33	3.4±0.1
1850°C-annealed SiC	AlLa5	1.8	4.4	514±18	4.9±0.3
	AlNd5	1.5	3.4	578±67	5.1±0.3
	AlY5	1.2	2.2	666±51	4.6±0.2
	AlYb5	2.4	5.7	718±42	3.9±0.1
1950°C-annealed SiC	AlLa5	3.2	6.7	659±55	5.5±0.3
	AlNd5	3.6	7.0	620±48	5.5±0.3
	AlY5	5.3	5.4	440±10	4.7±0.2
	AlYb5	5.0	4.6	371±35	4.9±0.1

Annealed at 1850°C, the grain growth in AlY5 was the smallest, and that in AlYb5 was the largest and most anisotropic among the four materials. The grain coarsening of all materials should have contributed to their increased fracture toughness.[12, 17] The toughening mechanism is thought to be the enhanced crack deflection and crack bridging by the large and elongated grains, as evidenced by the crack propagation features shown in Fig. 2. It is found that both the flexural

Silicon-Based Structural Ceramics

strength and fracture toughness of all materials improved after being annealed at 1850°C, in contrast to the well-known "trade-off" in improving toughness and strength of SiC ceramics, i.e., a coarse microstructure is beneficial for toughness but detrimental for strength, and *vice versa*.[9, 18] The improved flexural strength might be attributed to the interlocking force between the elongated grains and the effect of the increased toughness, under the condition that the grain sizes were not too large. In the four 1850°C-annealed materials, the largest mean grain size (of the annealed AlYb5) was just 2.4 μm.

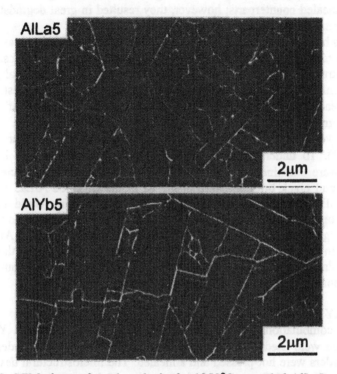

Fig. 2. SEM views of crack paths in the 1850°C-annealed AlLa5 and AlYb5 SiC materials.

With the annealing temperature increasing to 1950°C, AlY5 experienced the greatest grain growth and formed the coarsest microstructure among the four 1950°C-annealed materials. AlYb5 also underwent enormous grain growth, but the growth was less anisotropic than that during 1850°C-annealing; thus the 1950°C-annealed AlYb5 had a decreased mean aspect ratio. The mean grain size of the 1950°C-annealed AlLa5 and AlNd5 were much smaller, however their aspect ratios were the larger, indicating that the grain growth in these two materials was very anisotropic. Thus, the four 1950°C-annealed materials could be divided into two groups, with one including the annealed AlLa5 and AlNd5, and the other including the annealed AlY5 and AlYb5. The former had smaller grain sizes and larger aspect ratios than the latter. The relatively small grain sizes and large aspect ratios of the 1950°C-annealed AlLa5 and AlNd5 might account for the higher flexural strength and fracture toughness than their 1850°C-annealed counterparts. In the case of the 1950°C-annealed AlY5 and AlYb5, the coarser microstructures contributed to the higher fracture toughness than their 1850°C-annealed counterparts; however, they resulted in great degradation in the flexural strength.

It has been clearly shown above that structural parameters such as grain size and aspect ratio could affect the mechanical properties of SiC ceramics, however, the microstructure was not the only influencing factor. The mechanical properties were also influenced by the intergranular phase compositions. For instance, after being annealed at 1850°C, the annealed AlYb5 was composed of large and elongated grains, while the annealed AlLa5 had smaller grain size and aspect ratio; however, the latter had higher fracture toughness than the former (4.9 vs 3.9 MPa $\cdot$ m$^{1/2}$). To understand this, the crack propagation behaviors in the two materials were examined. As shown in Fig. 2, more frequently occurred crack deflection and crack bridging were observed in the annealed AlLa5 than in the annealed AlYb5. More crack deflection and crack bridging could result in a stronger toughening effect, which was thought to have contributed to the higher fracture toughness of the annealed AlLa5 than that of the annealed AlYb5. The difference in the crack propagation behaviors might be attributed to the different grain-boundary strength resulting from the different intergranular phase compositions.[1]

SUMMARY

Four combinations of rare-earth oxides (RE$_2$O$_3$; RE = La, Nd, Y and Yb) and alumina were used as sintering additives for a fine β–SiC powder, and the mixed powders were hot-pressed and annealed. The microstructural development during annealing and the mechanical properties of the annealed materials were different for the SiC doped with different RE$_2$O$_3$ additives. For the specimens AlLa5 and AlNd5, into which the added RE$_2$O$_3$ were La$_2$O$_3$ and Nd$_2$O$_3$, respectively, high temperature annealing treatment could lead to simultaneous improvement of flexural strength and fracture toughness. It is suggested that the mechanical properties of liquid-phase-sintered SiC can be improved to a higher level by optimizing the intergranular phase compositions and microstructures through judicious selection of sintering additives and annealing conditions.

ACKNOWLEDGMENTS
This work has been supported by METI, Japan, as part of the Synergy Ceramics Project. The authors are members of the Joint Research Consortium of Synergy Ceramics.

REFERENCES
(1) S. Prochazka, "The Role of Boron and Carbon in the Sintering of Silicon Carbide"; pp. 171-81 in *Special Ceramics 6*. Edited by P. Popper. British Ceramic Research Association, Stoke on Trent, U.K., 1975.
(2) H. Tanaka, "Sintering of Silicon Carbide"; pp. 213-38 in *Silicon Carbide Ceramics – 1*. Edited by S. Somiya and Y. Inomata. Elsevier, New York, 1991.
(3) L. S. Sigl and H. J. Kleebe, "Core-Rim Structure of Liquid-Phase-Sintered Silicon Carbide," *J. Am. Ceram. Soc.*, 76 [3] 773-76 (1993).
(4) N. P. Padture, "*In Situ*-Toughened Silicon Carbide," *J. Am. Ceram. Soc.*, 77 [2] 519-23 (1994).
(5) Y. –W. Kim, M. Mitomo and H. Hirotsuru, "Grain Growth and Fracture Toughness of Fine-Grained Silicon Carbide Ceramics," *J. Am. Ceram. Soc.*, 78 [11] 3145-48 (1995).
(6) J. J. Cao, W. J. MoberlyChan, L. C. De Jonghe, C. J. Gilbert and R. O. Ritchie, "*In Situ* Toughened Silicon Carbide with Al-B-C Additions," *J. Am. Ceram. Soc.*, 79 [2] 461-69 (1996).
(7) Y. Zhou, H. Tanaka, S. Otani and Y. Bando, "Low-Temperature Pressureless Sintering of α–Silicon Carbide with Al_4C_3-B_4C-C Additives," *J. Am. Ceram. Soc.*, 82 [8] 1959-64 (1999).
(8) G. –D. Zhan, M. Mitomo, H. Tanaka and Y. –W. Kim, "Effect of Annealing Conditions on Microstructural Development and Phase Transformation in Silicon Carbide," *J. Am. Ceram. Soc.*, 83 [6] 1369-74 (2000).
(9) D. Sciti, S. Guicciardi and A. Bellosi, "Effect of Annealing Treatments on Microstructure and Mechanical Properties of Liquid-Phase-Sintered Silicon Carbide," *J. Eur. Ceram. Soc.*, 21, 621-32 (2001).
(10) G. Rixecker, I. Wiedmann, A. Rosinus and F. Aldinger, "High-Temperature Effects in the Fracture Mechanical Behaviour of Silicon Carbide Liquid-Phase Sintered with AlN-Y_2O_3 Additives," *J. Eur. Ceram. Soc.*, 21, 1013-19 (2001).
(11) Y. Zhou, K. Hirao, M. Toriyama, Y. Yamauchi and S. Kanzaki, "Effects of Intergranular Phase Chemistry on the Microstructure and Mechanical Properties of Silicon Carbide Ceramics Densified with Rare-Earth Oxide and Alumina Additives," *J. Am. Ceram. Soc.*, 84 [7] 1642-44 (2001).
(12) P. F. Becher, "Microstructural Design of Toughened Ceramics," *J. Am. Ceram. Soc.*, 74 [2] 255-69 (1991).
(13) R. Chinn, "Grain Sizes of Ceramics by Automatic Image Analysis," *J. Am. Ceram. Soc.*, 77 [2] 589-92 (1994).

CORROSION OF SILICON NITRIDE MATERIALS IN ACIDIC AND BASIC SOLUTIONS AND UNDER HYDROTHERMAL CONDITIONS

M. Herrmann, J. Schilm, G. Michael
Fraunhofer-Institut für keramische Technologien und Sinterwerkstoffe, Dresden, Germany

ABSTRACT
The corrosion behaviour of gas pressure sintered silicon nitride ceramics with different amounts and compositions of the grain boundary phases was analysed in acids, bases and under hydrothermal conditions. The investigations show, that the corrosion is strongly affected by the SiO_2 content in the grain boundary phase and the amount of the additives. The differences in the corrosion kinetics are correlated with the structure of the formed corrosion layer. Some correlations between composition and corrosion stability in the different media are given.

1. INTRODUCTION
Silicon nitride-based ceramics are promising engineering materials for application under corrosive and wear conditions due to their hardness, mechanical strength and good corrosion resistance at room and elevated temperatures. However, during applications of silicon nitride ceramics in different media were found some degradations of the materials [1]. It was shown [1-5] that the materials exhibit nearly no degradation in acids (HCl, H_2SO_4 and HNO_3) at room temperature, but the grain boundaries of the silicon nitride materials dissolve in hot acids. The attack of acids in medium concentrations is much more pronounced than in diluted and in concentrated acids [1-6] (Fig. 1). The MgO-containing materials show better corrosion resistance than the Y_2O_3/Al_2O_3-containing materials. But up to now a lack of understanding of the corrosion behaviour exists [1,5,7]. Therefore based on experimental results and literature data basic ideas of the correlations between the composition of the materials and the corrosion behaviour of Si_3N_4 ceramics in different media are given in the paper.

2. EXPERIMENTAL
Silicon nitride ceramics of different compositions (Tab. 1) were prepared by gas pressure sintering. The details of the sample preparations are given in [5]. The materials were tested in 1 N H_2SO_4 in 1 N NaOH up to 130 °C and in distilled water up to 210 °C under pressure. The reaction vessels and sample holders were made of teflon. The acidic and basic solutions were stirred during the corrosion tests. The samples were rotated (300 rpm) during the hydrothermal corrosion tests. All solutions were regularly changed after 50 to 100 h. After corrosion the strength, the weight loss and the thickness of the corroded layer of the samples were measured. The surfaces of the materials were analysed using SEM. Cross-sections were analysed using light microscopy and SEM. More details of the sample preparation and conditions of the corrosion tests are given in [5, 8].

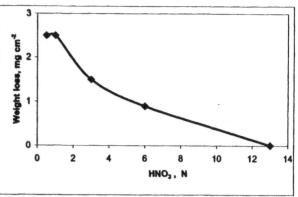

Fig. 1: Dependence of the weight loss of an Si_3N_4 ceramics with - MgO/ Al_2O_3 additive on the concentration boiling HNO_3 (200 h; after [6]).

Table I: Compositions of the investigated materials, strength (σ_0), residual strength (σ_r) and thickness of the corrosion layer after corrosion in 1N H_2SO_4 at 90 °C

Material	vol%	Additive	SiO_2 in GB [mass%]	σ_0 [MPa]	σ_r in H_2SO_4 [MPa][1]	Corrosion depth [µm][1]
AlMgSi 1	7.2	$MgAl_2O_4/SiO_2$	53	660		
AlMg 1	5.9	$MgAl_2O_4$	45	801	720	<5
MgAl 2	6.6	$MgAl_2O_4$	36	700	650	<10
MgAl 3	8.4	$MgAl_2O_4$	29	940	660	450
YAlMg 1	5.2	$MgAl_2O_4/Y_2O_3$	52	830	780	<5
YAlMg2	7.2	$MgAl_2O_4/Y_2O_3$	36	820		
YAl 1	10.9	Y_2O_3/Al_2O_3	20	920	470[2]	850[2]

SiO_2 in GB – SiO_2 content in the grain boundary phase, 1) after 700 h 2) after 300 h

Table. 2: Strength, residual strength σ (R)$_{4b}$, weight loss Δm and thickness of the corrosion layer for materials YAl1 and YalMg1 (200 h corrosion).

Sample	Strength σ$_{4b}$ [MPa]	1N NaOH (130°C)			20 % HF (20 °C)			1 N H$_2$SO$_4$ (boiling)			Hydrothermal corrosion (200 °C)		
		Δm, [mgcm²]	D [μm]	σ(R)$_{4b}$ [MPa]	Δm [mgcm²]	D [μm]	σ(R)$_{4b}$ [MPa]	Δm [mgcm²]	D [μm]	σ(R)$_{4b}$ [MPa]	Δm, [mgcm²]	D [μm]	σ(R)$_{4b}$ [MPa]
YAl-1	920±50	0.05	< 5	840±30	-0.4	20	440±10	-20.02	200	425±45	-0.19	<5	770±60
YAlMg2	820±100	0.17	30	740 ±30	-11	260	275±7	-0.35	<5	720 ± 60	-0.31	15	715±45

3. RESULTS

3.1 Stability in H$_2$SO$_4$

The results of investigations are given in **Table 1 and 2** and in **Figures 2 - 3**. With decreasing amount of sintering additives, the corrosion rate in H$_2$SO$_4$ at 90 °C was found to decrease very rapidly (**Fig. 2**). Analysis of corroded polished samples indicated that mainly the grain boundary phase in the triple junctions was attacked, whereas the grains and the thin grain boundary layers showed no signs of corrosion (**Fig. 3**). Also the residual strength of the materials whose corrosion depth took about 850 μm (YAl 1) showed values as high as 470 MPa indicating the low chemical attack of the grains and the thin grain boundary films between adjacent grains.

Fig. 2:
Dependence of the weight loss on the corrosion time in 1N H$_2$SO$_4$ at 90 °C:

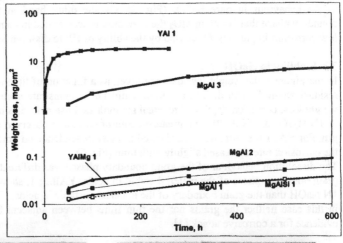

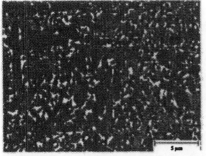 a)
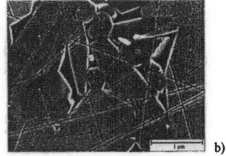 b)

Fig. 3:
SEM micrographs of the materials YAl
a) uncorroded surfaces; b) surfaces after 200 h in
1N H2SO4 at 90 °C and
c) surfaces after 50 h in 1N NaOH at 90 °C.

.c)

3.2 Stability in HF containing solutions.

Even at room temperature, the 20%-HF solution attacked the silicon nitride materials (Tab. 2). The HF solution attacked both the Si_3N_4 grains and the grain boundary phase resulting in a less stable corrosion layer. The corrosion resistance of the material, that was more stable in H_2SO_4 was lower in HF (Tab. 2). Additionally materials were tested in solutions of 1 N H_2SO_4 at 90 °C with different amounts of KF added (Fig. 4). The results show that small amounts of F⁻-additives in the corrosive media change the corrosion behaviour strongly. EDX measurements indicate that in the material containing a higher amount of Y_2O_3 as sintering additive, YF_3 is formed inside the corrosion layer, which reduces the corrosion in comparison to the behaviour in pure H_2SO_4 (Fig. 5). On the other hand in the material with grain boundary phase that is rich in SiO_2 the corrosion is accelerated in comparison to the relatively low corrosion in pure H_2SO_4 caused by the ability of HF to dissolve SiO_2.

3.3 Stability in NaOH

In the Figure 6 the weight loss data are given as a function of time and Table 2 presents the residual strength. The different materials did not demonstrate such extreme differences in weight loss, corrosion depth and residual strength as those materials that underwent corrosion in 1N H_2SO_4 did (Tab. 2). The greatest depth of corrosion observed after 200 h was only 30 μm. For YAl 1 the corrosion rate showed a linear dependence on time, whereas for YAlMg 1 the corrosion rate decreased slightly with time (Fig. 6).

The dissolution behaviour of the grain boundary phases was different from that in acids. The grain boundary phase of the material stable in acids, YAlMg 1, showed more degradation in 1N NaOH than the grain boundary of the material unstable in acids, YAl 1 did (Fig. 3c). Also in this case neither the grains nor the thin films between adjacent Si_3N_4 grains showed an evidence for a corrosive attack [5].

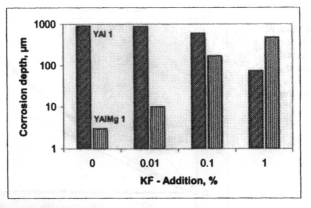

Fig. 4 Mass losses and corrosion depth of YAl 1 and YAlMg 1 corroded in 1N H_2SO_4 at 90°C for 100h with different amounts of KF added to the acid.

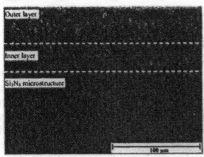
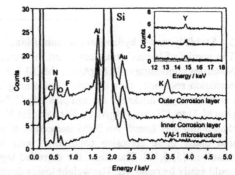

Fig. 5. SEM micrograph and EDX spectra of the corrosion layers and the Si_3N_4 microstructure of YAl 1 Peaks are labelled with the letters of the corresponding elements. (Au – was used for the coating)

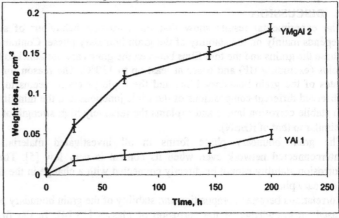

Fig. 6: Weight loss of Si_3N_4 ceramics YAl 1 and YAlMg 2 in boiling 1 N NaOH as a function of the time.

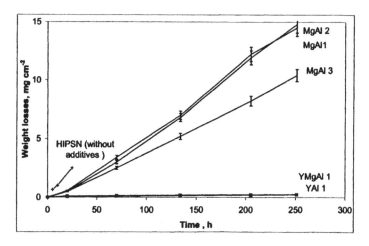

Fig. 7:
Weight losses of materials during hydrothermal corrosion at 210 °C (data for HIP SN taken from [7]).

3.4 Stability under hydrothermal conditions

The results of hydrothermal corrosion measurements are presented in Table 2 and Figure 7. Materials containing only $MgAl_2O_4$ (MgAl 1-3) as a sintering additive showed a much lower corrosion resistance than materials containing Y_2O_3 and $MgAl_2O_4$ or Y_2O_3 and Al_2O_3. The corrosion layer thickness for samples MgAl 1-3 after 250 h at 210°C ranged from 50 μm (MgAl 3) to 120 μm (MgAl 2). For samples YAl 1 a corroded layer with a thickness of less than 5 μm was found. An SEM inspection of the surfaces of the corroded samples showed that under these corrosive conditions, both Si_3N_4 grains and the grain boundary phase dissolved [5, 9,10]. Therefore, the corroded layer was unstable and thicker corrosion layers could easily be removed. The weight losses during corrosion depended linearly on time.

4. DISCUSSION

The experimental results show that the corrosion behaviour of silicon nitride ceramics depends mainly on the stability of the grain boundary phase. Contrary to the grain boundary phase the grains and the thin films between the grains are not attacked by a corrosive attack in acids (excluding HF) and bases at least up to 130°C. The reason for the different corrosion rates of the grain boundary films and the triple junctions seems to be connected with the observed different compositions of the triple junctions and the thin films [11,12]. This results in stabile corrosion layers and explains the relatively high strength of the corroded materials (similar to that of RBSN).

The grain boundary phase forms in all investigated materials a three-dimensional interconnected network even when its content is very low [5]. Therefore changes in the corrosion stability cannot be directly connected with a change in the distribution of the grain boundary phase.

Correlations between composition and stability of the grain boundary phases can be explained based on knowledge of the corrosion of glasses [13,14]. In the first step their corrosion behaviour in acids is featured by leaching of the network modifier (in our case Y, Mg and particularly Al) and formation of a hydrated glass network in the corroded regions. Schematically this can be illustrated by the following reaction:

$$\equiv(Si\text{-}O)_n\text{--}M + nH^+_{(aqu.)} \Rightarrow \equiv Si\text{-}OH + M^{n+}_{(aqu.)} \qquad (1)$$

For longer corrosion times, solution of the hydrated network is rate-controlling, as schematically shown in the reaction:

$$\equiv Si\text{-}O\text{-}SiO_{3-n}(OH)_n + H^+(aqu.) \Rightarrow \equiv Si\text{-}OH + SiO_{3-n}(OH)_{n+1}\,(aqu.) \qquad (2)$$

The stability of remaining hydrated network increases with decreasing amount of modifying ions, e.g. with increasing amount of bridging oxygen atoms in the network. Therefore the grain boundaries with a high content of SiO_2 are the most stable in acids. The introduction of one Mg^{2+} in the network results in 2 additional nonbridging oxygen, whereas the introduction of Y^{3+} results in the formation of 3 non bridging oxygen atoms., i.e. MgO additions result in less amounts of non bridging oxygen and higher stability. On the other hand during the sintering some of the MgO sintering additive usually evaporates by the reaction

$$6MgO + Si_3N_4 \Rightarrow 3 SiO_2 + 6 Mg + 2N_2 \qquad (3)$$

resulting in higher amount of bridging oxygen in the grain boundary phases and in higher stability. These changes of the compositions during sintering can result in higher scattering of the corrosion stability of materials with low additive contents.

Besides the overall similar tendencies in the stability of glasses and silicon nitride ceramics there exist differences. These differences are mostly connected with the formation of different corrosion layers. The partially hydrated glass network layer of the corroded glasses is not stable and can be removed easily. Therefore periodical ultrasonic treatments resulting in a change of the corrosion rate of oxide nitride glasses [5]. In the corroded Si_3N_4 ceramics this partially hydrated layer is stabilised by the strong Si_3N_4 skeleton. An evidence for this is that the starting period of the corrosion kinetics of Si_3N_4 ceramics (up to several 10 h) can be described by the corrosion behaviour of oxide nitride glasses determined under conditions without ultrasonic treatment [5]. For higher corrosion depth the corrosion rate of the ceramics decreases rapidly. The reaction control of the process observed for the beginning of the corrosion change to diffusion controlled mechanisms and finally reaches a nearly complete passivization. The kinetics of the corrosion are explained more in detail in [5, 8]. It can be strongly assumed that the network of the corroded, porous Si_3N_4 microstructure reduces the diffusion of the SiO_2 x H_2O structures from the reaction front inside the ceramic material to the outer surface and force the condensation reaction of the SiO_2 x H_2O structures inside the corrosion layer, which in turn lead to a strongly decreased solubility of the partial hydrated SiO_2 network [5]. It reduces the further destruction of the network and the leaching of the network modifiers. The formation of a stable passivizating SiO_2 network in the ceramics is strongly accelerated by an increasing amount of SiO_2 in the grain boundary phases of the ceramics, by reducing the size of the triple junctions and by reduction of the additive content [8]. With increasing corrosion temperature the passivization increases (at least up to 90 °C) [8]. This mechanism also seems to be the reason for the high stability of silicon nitride ceramics in concentrated acids [1,6]. Also the oxidation of the surface prior to corrosion increases the corrosion resistance drastically and can be explained by this model. Recently microanalysis of the corroded layers showed a gradient in the concentration of the oxygen content, indicating the formation of such layers [15]. The assumption of the formation of a stable hydrated SiO_2-rich network is also in agreement with the different corrosion behaviours in HF, in bases and under hydrothermal conditions.

Since Fluorine ions are known to dissolve the SiO_2 glass network by the reaction

$$6 HF^- + SiO_2 \longrightarrow H_2[SiF_6] + 2 H_2O$$

the materials with a high SiO_2 content in the grain boundary phase lose there stability in acids in the presence of F^- ions. Also the corrosion layer is less strong as it is in pure H_2SO_4 due to the attack of the grains and the thin films between the grains. This results in the low strength of the materials in HF (Tab. 2).

In NaOH the materials containing MgO and high amounts of SiO_2 exhibit a lower stability than materials with a grain boundary phase consisting of yttria and alumina. On the one side this is connected with the low solubility of $Y(OH)_3$ formed under these conditions. On the other side the SiO_2 network starts to dissolve and therefore grain boundary phases rich in SiO_2 have a lower stability.

Silicon-Based Structural Ceramics

Under hydrothermal conditions SiO_2-rich glasses are less stable than alumina and yttria rich glasses. Therefore, the thin films between the grains also dissolve and, as a consequence, under hydrothermal conditions unstable corrosion layers result [5,9]. Additionally, the grains start to dissolve at temperatures above 200 °C in H_2O [5, 7, 9]. These processes are connected with the higher solubility of SiO_2 under these conditions. This is also the reason, why additive-free HIPSN with only about 3 wt% SiO_2 in the grain boundaries shows a much higher corrosion rate than Y_2O_3/Al_2O_3-containing materials do. Under hydrothermal conditions mostly a linear corrosion kinetic can be found; i.e., the formation of pronounced protective layers is not observed. A higher stability of the yttria and alumina rich grain boundaries than the Si_3N_4 grains was observed during corrosion at 270 °C [7,9]. The MgO-containing materials Mg Al 1-3 have a similarly high corrosion rate as the HIPSN. Whereas the dissolution rates of the grains and the grain boundary phases for YAl 1 at 210 °C are similar, the grain boundary phases in the materials MgAl 1-3 dissolve much faster than the grains, causing a very weak corrosion layer to be formed.

Based on the experimental and literature data the corrosion behaviour can be schematically summarised as shown in Fig. 8 and 9.

5. CONCLUSIONS

The stability of silicon nitride materials in acids and bases and under hydrothermal conditions strongly depends on the amount and composition of the grain boundary phase.

Silicon nitride materials with a high SiO_2 content in the grain boundary phase are more stable in acids and less stable in bases and under hydrothermal conditions. Materials with a low SiO_2 content in the grain boundary phase are more stable in bases and under hydrothermal conditions and less stable in acids. This behaviour is more pronounced in materials with a higher amount of grain boundary phase.

The results of our investigations indicate that tailoring of the microstructure and the grain boundary composition of silicon nitride ceramics is necessary if these materials have to be applied in corrosive environments.

By changing the composition of the material, the corrosion resistance (weight loss and thickness of the corrosion layer) can be changed by more than two orders of magnitude and the reduction of strength minimised.

6. ACKNOWLEDGEMENTS

This study was supported by the DFG under contract number HE 2471/1 and by the BMWi and the AiF under contract Nr. 12130 BR.

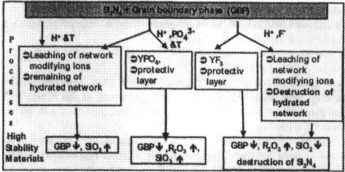

Fig. 8: Schematic presentation of the processes taking place during corrosion of Si_3N_4 ceramics in different acids

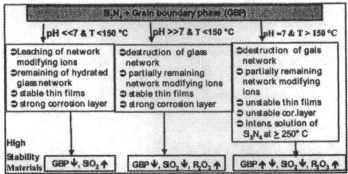

Fig 9: Schematic presentation of the processes taking place during corrosion of Si₃N₄ ceramics in different media.

7. LITERATURE

1. Petzow G., Herrmann M.; Silicon Nitride Ceramics. In: Structure and Bond, Vol.102 ,(2002), Springer, Berlin, 47-166
2. Komeya K, Meguro T, Atago S, Lin CH, Abe Y, Komatsu M (1999) Corrosion Resistance of Silicon Nitride Ceramics. In: Niihara K, Sekino T, Yasuda E, Sasa T (eds) The Science of Engineering Ceramics II. Key Eng.Mat. 161-163, Trans Tech Publications, Switzerland, p 235
3. Okada A., Yoshimura M., Mechanical degradation of Silicon nitride ceramics in corrosive solutiona of boiling sulphuric acid, Key Engineering Materials, 113 (1996), 227 – 236
4. Monteferde F., Mingazzini C., Giorgi M., Bellosi A.; Corrosion of silicon nitride in sulphuric acid aqueous solution, Corrosion Science 43 (2001), 1851 – 1863
5. Herrmann M., Schilm J., Michael G., Meinhard J., Flegler R.; Corrosion of Silicon Nitride Materials in Acidic and Basic Solutions and under Hydrothermal Conditions J. Europ. Ceram. Soc. 2002 in print
6. Kanbara K., Uchida N.,. Uematsu K., et all.; Corrosion of Silicon Nitride Ceramics by Nitridic Acid , Mat. Res. Soc. Proc. Vol. 287, (1993) 533-538
7. Herrmann M., Krell A., Adler J., Wötting G., Hollstein T., Pfeifer W. , Rombach M.; Keramische Wälzlager für Anwendungen in korrosiven Medien, VDI-Ber. 1331 (Innovationen fuer Gleitlager, Waelzlager, Dichtungen und Fuehrungen), (1997), 251-258
7. Schilm J, Herrmann M. , Michael G.; Kinetic study of corrosion of silicon nitride materials in acids, J. Europ. Cer. Soc. (in print)
9. Herrmann M., Schilm J., Michael G., Adler. J.; submitted to cfi 2002
10. Schilm J., Herrmann M., Michael G.; submitted to J. Europ. Ceram. Soc. (2002)
11. Bando Y., Mitomo M., Kurashima K. J.; An inhomogeneous Grain Boundary Composition in Silicon Nitride, Mater. Synthesis and Processing 6, (1998), 359
12. Bobeth M., Clarke D.R., Pompe W.; J. Am. Ceram. Soc. 82, (1999), 1537
13. Paul A. Chemical Durability of glasses; a thermodynamic approach, J. Mater. Sci. 12, (1977), 2246-2268
14. White W.B. (1992) Theory of Corrosion of Glass and Ceramics. In: Clark D.E., Zoitos B.K. (eds) Corrosion of Glass, Ceramics and Ceramic Superconductors. Noyes Publications , USA,
15. B. Seipel B., Nickel K. G, J. Europ. Cer. Soc. (2002) (in press)

Applications

DEVELOPMENT OF HIGH-TEMPERATURE HEAT EXCHANGERS USING SIC MICROCHANNELS

Merrill A Wilson, Raymond Cutler,
Marc Flinders, Matt Quist, Darin Ray
Ceramatec Inc.
2425 S. 900 West
Salt Lake City, UT 84119

Dr. Darryl Butt, KY. Kim,
E. L. Pabit
University of Florida
PO Box 116400
Gainesville, FL 32611

ABSTRACT

Efficiency and emissions of advanced gas turbine power cycles can be improved by incorporating high-temperature, natural gas fired ceramic heat exchangers. In cooperation with the DOE, preliminary development and testing of SiC based structures has been completed. This program has focused on three initial areas: thermo-mechanical degradation as a function of operating environments, design of layered composite porous/dense micro-channel devices, and fabrication development of these structures. A summary of the progress and findings will be discussed.

INTRODUCTION

High efficiency in power generations systems has driven by several factors, such as fuel utilization, operating costs and emissions. Fundamentally, all heat cycle efficiencies are limited by the maximum operating temperature for the materials of construction. Continued engineering improvements will likely result in incremental performance improvements and slightly lower emissions, but step improvements will result only from step changes in materials technology. Introduction of high temperature ceramics will redefine the operating window and provides step improvements in efficiency. Figure 1 is an adaptation of Lee et al.[1] which summarizes the power cycle efficiency as a function of rotor inlet temperature. Overlaid onto this plot are the temperature limits of various heat exchanger materials.

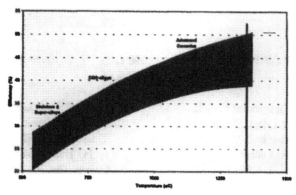

Figure 1. Power Cycle Efficiency as a Function of Rotor Inlet Temperature

This research focuses on the development of a ceramic micro-channel heat exchanger. The basic design of this heat exchanger incorporates the high temperature properties of Silicon Carbide (SiC) and the process intensification of micro-reactors. SiC has been shown to perform well in direct fired heat exchanger applications[2]. The dimensional scales found in micro-reactors reduce the spatial resistances for heat and mass transfer and thus minimize the time and volume required for unit operations. These two technologies are integrated into thin, internally channeled heat fins where natural gas is injected into a cross-flowing air stream. Conceptually, a series of heat fins are stacked into a modular unit separated by flow spacers. Figure 2 illustrates the concept of a high-temperature ceramic heat exchanger.

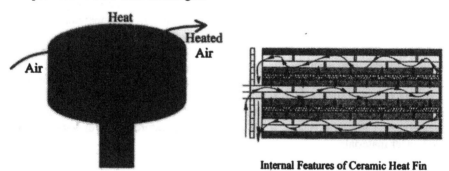

Internal Features of Ceramic Heat Fin
Figure 2. Conceptual Design of a Modular Heat Fin Stack

MATERIALS THERMO-MECHANICAL PERFORMANCE

Thermodynamics of SiC Corrosion

At the operating conditions, the high temperature heat exchanger will be exposed to the natural gas fuel stream and the combustion product stream. Based on the results of the thermodynamic calculations (FACTSAGE equilibrium software and SGTE database), and published literature[3,4,5,6], the degradation of SiC will occur due to $SiO_2(s)$ formation and subsequent volatilization as $SiO(OH)_x(g)$ or $Si(OH)_x(g)$ in the presence of $H_2O(g)$.

Experimental Set-up and Procedure

Gas Selection: The thermodynamic calculations were used to determine an analog gas mixture for the heat treatment and durability studies; this mixture is found in Table I. This composition was selected to simulate the indirect fired conditions and avoid carbon deposition on the SiC materials. Although the feed mixture did not contain H_2O, at equilibrium test conditions (1200°C, 1300°C and 1350°C) the water content would be about 3.5%. Thus, $H_2O(g)$ was present to simulate the oxidation of SiC and volatilization of $SiO(g)$.

Table I. *Test gas composition and its equilibrium composition at 1200°C, 1300°C and 1350°C.*

Gases	Nominal Composition	Equilibrium amount (atm)		
		@1250°C	@ 1300°C	@ 1350°C
CH_4	0.3	2.29×10^{-7}	1.27×10^{-7}	7.35×10^{-8}
H_2	27.3	2.43×10^{-1}	2.43×10^{-1}	2.43×10^{-1}
N_2	55.3	5.50×10^{-1}	5.50×10^{-1}	5.50×10^{-1}
CO	12.5	1.64×10^{-1}	1.65×10^{-1}	1.65×10^{-1}
CO_2	4.6	8.70×10^{-3}	8.22×10^{-3}	7.80×10^{-3}
H_2O	0.0	3.40×10^{-2}	3.45×10^{-2}	3.49×10^{-2}

Exposure Experiment: Rectangular samples of SiC with dimensions of 48.5 mm x 4.4 mm x 4.4 mm (LxWxH), and boron-doped SiC (Hexoloy SA, Carborundum Co.), with dimensions of 44.7 mm x 4.0 mm x 3.0 mm, were used for the durability studies. Additionally, SiC powder (3.2 g/cm^3 and 0.27μ grain size) samples were also used to assess the oxidation kinetics.

Exposures samples were placed in a high temperature furnace with the tensile side of the specimen were facing upward. SiC powder samples in a 1½" by 1½" boat were also simultaneously exposed. Prior to exposure testing, the furnace was evacuated and purged with argon gas. As the exposure temperature was reached, the samples were introduced and maintained in the mixed gas atmosphere.

Thermogravimetric analysis of SiC powder samples (20 mg) were loaded into a TGA, purged with argon and then brought to the desired temperature. When the set temperature was reached, the mixed gas was introduced. The system was then maintained at the set temperature and exposure for 12 hours.

As-received and heat treated bend bars were fractured at room temperature using an MTS ASTM D790 four-point bend test at a loading rate of 0.05 cm/min. All samples were tested immediately after heat treatment.

Fractography via scanning electron microscopy (SEM) was used to identify the sites of failure initiation in the mechanical test specimen. In most cases, classic mirror, mist, and hackle regions were easily visible.

Results and Discussion

Heat treatment and durability studies of SiC materials in the simulated gas mixture were performed. All of the samples showed weight gains and room temperature strength increases when compared to the as-received materials. Simultaneous changes were observed in the evolution from the platelet-like grains of the as-received materials to a more continuous surface in the exposed samples, as shown in Figures 3a and 3b. EDX of the heat-treated samples showed the presence of oxygen peaks, suggesting formation of SiO_2 in the specimens.

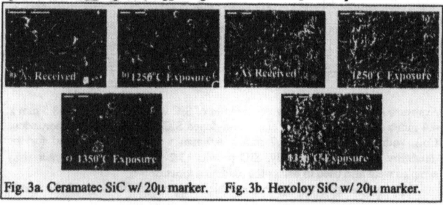

Fig. 3a. Ceramatec SiC w/ 20µ marker.   Fig. 3b. Hexoloy SiC w/ 20µ marker.

Kinetics of SiC in Fuel Mixture: A summary of the weight change observed for the monolithic SiC and powder samples are shown in Figures 4. The weight gains for the 36 hour exposure Ceramatec and Hexoloy materials exhibited similar behaviors; both materials increased up to 1300°C (SiO_2 formation) and decreased at 1350°C (SiO volatilization). The Hexoloy weight gain exceeded that of the Ceramatec SiC; this suggested that the Ceramatec material is more oxidation resistant in the simulated environment. A possible explanation for the lower weight gain in the powder sample may be that the two monolithic materials have some impurities, which were introduced during processing. It was reported[7] that presence of impurities increases the oxidation rate of SiC materials.

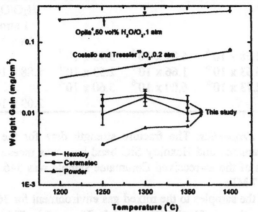

Fig. 4. Comparison of the weight gain observed in our experiment with that observed by Opila[8] and Costello and Tressler[9].

Weight gain observed for exposure time of 36 hours at different temperature has been compared to works found in literature. Data shown for Opila[8] was based on parabolic rate constant for oxidation of SiC in 50-vol% H_2O/O_2 mixtures at a total pressure of 1 atm. The data of Costello and Tressler[9] was based on parabolic rate constants for the oxidation of sintered α-SiC doped with boron, similar to the Hexoloy material, in a reduced O_2 pressure of 0.2 atm for 24 hours. It can be seen in Table II that the oxidation observed in the indirect fired simulant is order of magnitude lower than that observed in 0.2 atm O_2[9] and 50-vol% H_2O/O_2[8].

Table II. *Comparison of parabolic rate constant of the SiC material.*

		Parabolic rate constant, K_{par} (mg^2/cm^4-h)			
Temp.(°C)	Ceramatec	Hexoloy	Powder	Opila[8] 50-vol% H$_2$O/O$_2$, 1 atm	Costello and Tressler[9] O$_2$, 0.2 atm
1200				1.18×10^{-3}	
1250	2.58×10^{-6}	6.87×10^{-6}	3.75×10^{-7}		
1300	6.63×10^{-6}	1.66×10^{-5}	2.53×10^{-7}	1.88×10^{-3}	6.21×10^{-5}
1350	2.73×10^{-6}	6.00×10^{-6}	5.60×10^{-6}		
1400				2.09×10^{-3}	1.68×10^{-4}

Mechanical Properties: The fracture strength data for the as-received and heat-treated Ceramatec and Hexoloy SiC bend bars was measured. The average fracture strength of the as-received Ceramatec sample was 346 MPa and that of Hexoloy was 297 MPa.

Exposure of the samples to the mixed gas environment for 36 hours increased the fracture strength, σ_f, of both SiC materials (Figure 5a). This can be attributed to the formation of a continuous surface from the platelet-like grains of the as-received samples. The formation of continuous surface and the reduction of failures due to surface cracks, suggest that surface crack healing occurred, which may also have contributed to the increase in fracture strength of the heat-treated SiC materials. It is also apparent that Ceramatec SiC has a higher comparative strength than the Hexoloy materials at all the exposure temperatures tested.

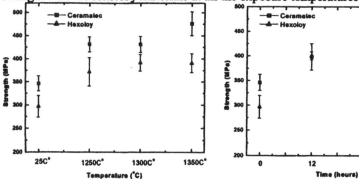

Fig. 5a. Plot of room-temperature bend bar strength of SiC materials exposed for 36 hours.

Fig. 5b. Plot of room-temperature bend bar strength of SiC materials exposed at 1300°C.

Silicon-Based Structural Ceramics

Strength versus time profile for the SiC materials exposed at 1300°C is shown in Figure 5b. The strength of both materials increased with exposure time. However, there was a reduction in the strength of the Hexoloy material after 100 hours of exposure, and was statistically equivalent to that of the as-received material. It is also apparent from the figure that the Ceramatec SiC has a higher strength after 100 hrs of exposure, as compared to the Hexoloy materials.

COMPONENT ENGINEERING AND FABRICATION
Heat Fin Design
As seen in Figure 2, the heat fin structure is an assembly of multiple layers with the functions of flow distribution, flow regulation, gas mixing, oxidation control and gas stream isolation. The prototypical heat fins designed and fabricated have symmetric structures consisting of 14 individual layers of approximately 200μ each. These layers are illustrated in Figure 6.

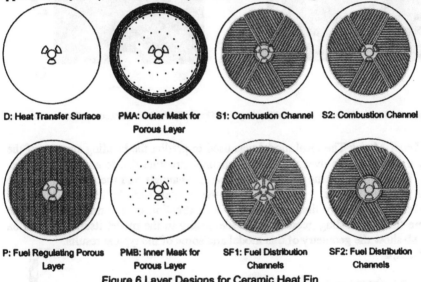

D: Heat Transfer Surface   PMA: Outer Mask for   S1: Combustion Channel   S2: Combustion Channel
                            Porous Layer

P: Fuel Regulating Porous   PMB: Inner Mask for   SF1: Fuel Distribution   SF2: Fuel Distribution
Layer   Porous Layer   Channels   Channels

Figure 6 Layer Designs for Ceramic Heat Fin

Finite Element Analysis
In order to determine the thermo-mechanical integrity of the proposed heat fin, two Finite Element Models (FEM) were developed to predict the global stresses and the local stresses. The global model was used to determine the stresses from macroscopic boundary conditions and loads, such as cross-flow convection and internal combustion. The local model included the detailed internal features to

determine the bending and stresses due to external pressures on thin membranes that span the internal channels.

Global Model: The global model provided the thermo-mechanical response to the internal oxidation/combustion of the fuel for air preheat. This model was used to establish the allowable ceramic to air thermal gradient criteria by coupling a thermal analysis (TOPAZ3d) with a mechanical analysis (NIKE3d).

The global model was used to determine the cooling effect of the external air and the internal fuel oxidation on the overall component stresses. This model also migrated to a "square" design for improved packing density for overall system design. Typical results from the thermal model and the coupled mechanical stress model are found in Figures 7a and 7b.

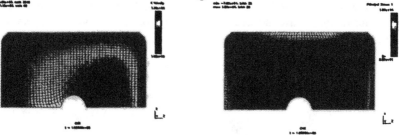

Figure 7a Thermal Model – External Cooling   Figure 7b Mechanical Model – External Cooling

Local Model: The local model was used to predict the bending stresses of the heat transfer surface over the internal channel structure due to external pressure. This model included the geometry of several channels and the external ceramic heat transfer surface. The edges of the model had "reflective" boundary conditions that imply that the model is an extraction of a common repeat unit where the surrounding geometry acts identically to the model itself. Figures 8a and 8b show the geometry of this model and some of the typical results.

Figure 8a Local Model - Geometry   Figure 8b Local Model – Typical Results

From these results, it was found that the local bending stresses are secondary to the global stresses due to thermo-mechanical loading. The local model included several design variables wherein sensitivities could be analyzed to formulate design rules (see Figure 9). These variables included the flow channel width and depth, channel wall width, channel skew angle, thickness of the heat transfer layer and the applied pressure load. The sensitivity studies indicate that the width of the channel has the strongest effect on the local stresses.

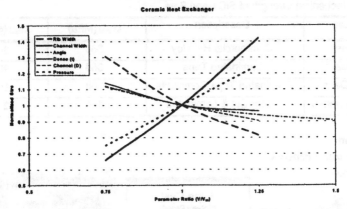

Figure 9 Design Rules for Local Stress Design

Featured Bend Bars: The mechanical strength of the SiC was assessed using 4 point bend bars. Early baseline properties were obtained from bar samples made from commercially available SiC wafers and cut into test specimens. Additional strength samples were fabricated from tape cast layers processed identically to the heat fin structures. Additionally, bend bars with porous or featured layers on the tensile surface were made to assess the degraded strength properties due to pores and slots (channels). Figure 10 illustrates these "modified" bend bars and Table III gives the strength results for the various samples tested.

Figure 10 Modified SiC Strength Bend Bars

It was found that the average strength of the dry pressed bars and the tape cast bars were quite similar. However, the strength for the modified bars exhibited a lower strength, as would be expected. This is due to the increase the number of crack initiation sites (surfaces of pores and slots) and stress intensifiers resulting from the reduced area (30% porous or void fraction of slotted layers) and corners of these modified samples. These reduced strength values determined the design stresses for the various layers within the heat fin.

Table III Mechanical Strength of SiC Bend Bars

Sample	Description	σ_{ave} (MPa)	σ_{SD} (MPa)
Dense	Commercial Hexaloy	360	30
Dense	Laminated Tape	368	90
Porous/Dense	Laminated Tape Composite	268	19
Featured/Dense	Laser Cut & Laminated	164	27

Fabrication
Fabrication Process:

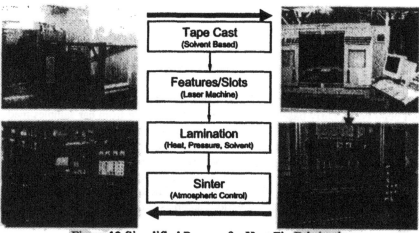

Figure 12 Simplified Process for Heat Fin Fabrication

Prototypical Structures: The initial process development was done using Al_2O_3 based ceramic tapes and as these processes matured, were migrated to SiC based tapes. Several of the Al_2O_3 and SiC based micro-channel heat fin structures were fabricated. Photos of these components are found in Figure 13.

Silicon-Based Structural Ceramics

Green (unfired) Bodies

Al₂O₃ Micro-Channel Structure

SiC Heat Fin Internal Features   **SiC Heat Fin Structures**

Knitting of Dense and Porous Layers   **Morphology of Porous Layers**

Figure 13 Al₂O₃ and SiC Micro-Channel Heat Fin Structures

ACKNOWLEDGEMENTS

This work was done with funding provided by the Department of Energy – NETL Small Business Innovative Research Program (DOE Grant Number: DE-FG 03-01ER83211).

REFERENCES

[1] Lee, J. C., Campbell, J., Jr., and Wright, D. E., "Closed-Cycle Gas Turbine Working Fluids," Presented at the *ASME Gas Turbine and Products Show*, New Orleans, La, March 1980; also *J. Engg. for Power*, 220-228 (1981).

[2] T.Darroudi, J.R. Hellmann, R.E. Tressler, and L. Gorski, "Strength Evaluation of Reaction-Bonded Silicon Carbide Radiant Tubes after Long-Term Exposure to Combustion and Endothermic Gas Ambients," *J. Am. Ceram. Soc.*, 1992, [Vol. 75 no.12] (pp. 3441-3451).

[3] D. E. Fox, E. J. Opila, and R. E. Hann, "Paralinear Oxidation of CVD SiC in Simulated Fuel-Rich Combustion," *J. Am. Ceram. Soc.*, 83 [7] 1761-67 (2000).

[4] E. J. Opila, J. L. Smialek, R. C. Robinson, D. S. Fox, and N. S. Jacobson, "SiC Recession Caused by SiO_2 Scale Volatility under Combustion Conditions: II, Thermodynamics and Gaseous-Diffusion Model," *J. Am. Ceram. Soc.*, 82 [7] 1826-34 (1999).

[5] S. T. Tso and J. A. Pask, "Reaction of Fused Silica with Hydrogen Gas," *J. Am. Ceram. Soc.*, 65 [9] 457-60 (1982).

[6] T. Narushima, T. Goto, Y. Yokoyama, Y. Iguchi, and T. Hirat, "High Temperature Active Oxidation of Chemically Vapor-Deposited Silicon Carbide in $CO-CO_2$ Atmosphere," *J. Am. Ceram. Soc.*, 76 [10] 2521-24 (1993).

[7] T. Narushima, T. Goto, T. Hirai and Y. Iguchi, "High-Temperature Oxidation of Silicon Carbide and Silicon Nitride," *J. Am. Ceram. Soc.*, 74 [10] 2583-2586 (1991).

[8] Elizabeth J. Opila, "Variation of the Oxidation Rate of Silicon Carbide with Water-Vapor Pressure," *J. Am. Ceram. Soc.*, 82 [3] 625-36 (1999).

[9] J. A. Costello and R. E. Tressler, "Oxidation Kinetics of Silicon Carbide Crystals and Ceramics: I, in Dry Oxygen," *J. Am. Ceram. Soc.*, 69 [9] 674-81 (1986).

CHARACTERIZATION OF CERAMIC COMPONENTS EXPOSED IN INDUSTRIAL GAS TURBINES

M. K. Ferber and H. T. Lin
Metals and Ceramics Division
Oak Ridge National Laboratory
Oak Ridge, TN 37831-6068

ABSTRACT
This paper provides a review of recent studies undertaken to examine the mechanical and thermal stability of silicon nitride ceramic components that are currently being considered for use in gas turbine applications. Specific components examined included a bowed ceramic nozzle evaluated in an engine test stand, ceramic vanes exposed in two field tests, and an air-cooled vane that is currently under development. Scanning electron microscopy was used to elucidate the changes in the microstructures arising from the environmental effects. The recession of the airfoils resulting from the volatilization of the normally protective silica layer was also measured. The stability of the intergranular phases was evaluated using x-ray diffraction. The surface strength was measured using a miniature biaxial test specimen, which was prepared by diamond core drilling.

INTRODUCTION
Over the last 30 years, a number of programs in the United States have sought to introduce monolithic ceramic components into gas turbines with the goals of increasing efficiency and lowering emissions. High performance silicon nitride and silicon carbide ceramics typically have been leading candidates for use in these applications. In spite of their potential, ceramic components exposed in engine tests [1-3] have not exhibited the required reliability. One factor for this limited success is that the test specimens, which are used to establish a database for life prediction, are generally fabricated differently from the turbine component. Consequently the properties measured in the laboratory may not reflect those of the component. For example, surface fracture may be more frequent in the

specimens with relatively high surface area to volume ratios, as compared with that of an engine component. The reverse may be true as far as volume fracture is concerned. The component may also experience a changing surface flaw population (due to contact damage, particle impact, creep, etc.), which is difficult to reproduce in a laboratory environment. Furthermore, laboratory specimens are generally machined to maintain the tight dimensional tolerances required for a valid test. However, components (particularly airfoils) contain as-processed surfaces, which may exhibit dramatically different properties compared with the bulk material. For thin sections these properties will control the overall behavior of the component. In the case of ceramic materials, processing (green-state forming and high-temperature densification) of the components and test specimens may be different leading to differences in the mechanical performance (Figure 1).

The second factor involves the deleterious effects of the gas turbine environment upon component lifetime. In such environments, three processes can impact performance of the silicon nitride: (1) localized corrosion due to the presence of reactive species in the environment, (2) environmentally-induced destabilization of the intergranular phases, and (3) rapid recession of the silicon nitride (or silicon carbide) due to loss of the silica scale by direct reaction with water vapor. Localized corrosion can occur when metallic impurities such as iron or nickel are deposited on the airfoil surface. The oxidation of these impurities disrupts the silica scale leading to the formation of a surface pit.

Environmental attack can also result in subsurface phase transformations that give rise to stresses large enough to nucleate cracks [4].

Oxidation of both silicon nitride and silicon carbide is increased by (1) the replacement of oxygen by water vapor and (2) an increase in the pressure of the oxidant. In addition, the high velocities and presence of water in the environment can lead to the volatilization of the normally protective silica layer. Researchers at NASA Glenn Research Center [5-7] have shown that the presence of water vapor leads to the formation of a gaseous $Si(OH)_4$ species via a reaction with the silica layer. The rate of formation, k, of this species and thus the rate at which the SiC (or Si_3N_4) is consumed by its continued oxidation is determined by the expression,

$$k \, \mu \, v^{1/2} \, P(H_2O)^2/(P_{total})^{1/2} \qquad (1)$$

where v is the gas velocity, $P(H_2O)$ is the pressure of the water vapor, and P_{total} is the total pressure. For lean burn conditions, k for a number of SiC based materials is found to be well described by the expression,

Silicon-Based Structural Ceramics

$$k \, (mg/cm^2h) = 2.04 \, \exp(-108 \, kJ/mole/RT) \, v^{1/2} \, P_{total}^{3/2} \qquad (2)$$

where T is the absolute temperature (K), P_{total} is the total pressure (atm), and v is the velocity (m/s). In addition to the volatility issues, which may lead to loss of function due to excessive changes in component dimensions, the effect of the environment upon the mechanical stability must be understood as well

A major issue with these environmental effects is that they are not easily reproduced in the laboratory. Although it is possible to conduct mechanical tests under conditions of high-pressure and high-temperature water vapor, representative gas turbine velocities cannot be obtained. Burner rigs are capable of generating both high-pressure and high-velocities but they generally do not have provisions for applying controlled mechanical stresses.

One novel approach for avoiding the problems arising from material variations (between standard test specimens and the component) is to evaluate the component properties directly using small, specialized test specimens. For curved airfoil surfaces, miniature disks, which are prepared by diamond core drilling and back machining, can be used to quantify retained surface strength. Sectioning of components that have been exposed in test stands or field tests also provides for quantifying time-dependent changes of a number of other thermal and mechanical properties. Extensive microstructural evaluation can be used to further define the effect of environment and service conditions on mechanical reliability. This paper summarizes results of component characterization studies implemented in support of recent programs at Solar Turbines, Rolls-Royce, and Pratt & Whitney.

RESULTS AND DICUSSION
Solar Turbines

Solar Turbines was awarded the Ceramic Stationary Gas Turbine (CSGT) Development contract from the Department of Energy (DOE) in 1992 [8]. The goals of this program, which continued through 2000, were to improve engine performance (fuel efficiency, output power) and reduce exhaust emissions. The approach involved retrofitting an existing gas turbine (Solar's Centaur 50S) with silicon nitride nozzles (vanes), silicon nitride blades, and SiC-based composite combustor liner.

In September 1998, a 100-hour nozzle engine test was initiated in which the engine was subjected to cold and hot engine restarts and shutdown cycles that progressively increased in severity. The nozzles were fabricated from SN88 silicon nitride manufactured by NGK Insulators, Ltd. Borescope inspections, which were conducted after shutdown cycles, revealed cracking after 68 hours (Figure 2). Selected vanes were sectioned for detailed scanning electron microscopy (SEM). As show in Figure 3, a light-colored region or oxidation zone was found to have evolved in the vicinity of the primary crack. Subsequent indentation toughness measurements revealed that this zone exhibited a much lower fracture toughness compared with the bulk material. The asymmetry of the cracks extending from the corners of the Vickers indenter further suggested the presence of residual stress. Both of these observations may have been a factor in the formation of the microcracks along the airfoil surface. X-ray diffraction analyses indicated that changes in the microstructure and secondary phase in the oxidation zone were due to the transformation of the intergranular phase (J-phase – $Yb_4Si_2O_7N_2$) to $Yb_2Si_2O_7$ plus $Yb_2Si_2O_5$. The volume decrease for this transformation was responsible for the residual stress and reduction in fracture toughness. Based upon correlations of the fracture locations with the predicted temperature distributions, it was also established that formation of the oxidation region occurred at a relatively low temperature (about 850°C). Subsequent laboratory studies [4] showed that this behavior occurred as a result of the loss of the normally protective silica scale. This intermediate temperature effect was responsible for a significant slow crack growth susceptibility.

Rolls-Royce

The vanes examined in this program [2,9] were fabricated from either AS800 or SN282 silicon nitride. Both materials, which are densified using rare-earth sintering aids, are classified as self-reinforced. The manufacturers of the AS800 and SN282 are Honeywell Ceramic Components in Torrance, CA and Kyocera Ceramic Components in Vancouver, WA.

The first stage ceramic vanes (Figure 4) were designed for retrofit into a Model 501-K turbine (Rolls-Royce Allison, Indianapolis, IN). Following a 22 h shakedown run in a test turbine, the first stage vane assembly was mounted in a Model 501-K turbine at a commercial site (Exxon - Mobile, AL). During the field tests at Exxon, the average temperature and pressure of gas entering the vanes were approximately 1066°C (1950°F) and 8.9 atm (128 psia), respectively. However, due to the combustor temperature pattern, the mid-span gas temperature could have been as high as 1288°C (2350°F) at the "hot spot". The inlet gas velocity at vane mid-span was about 162 m/s (530 ft/sec) and the gas accelerated to about 573 m/s (1880 ft/sec) at the vane exit. Due to the extremely humid conditions during the test, the mole fraction of water vapor for the gas entering the vanes was calculated to be 0.101.

Two separate field trials were conducted. In Test 1, only uncoated AS800 vanes were evaluated. Although no vane failures occurred during the field test, dimensional measurements indicated that recession of the silicon nitride was a problem. Consequently, the engine test was terminated after 793 h (815 total time including shakedown test).

Figure 5 summarizes the micrometer-based thickness measurements obtained for the mid-span region of the trailing edge (symbols). Each point represents a separate vane. The loss of material, which was excessive for a number of vanes, is due to the inability of the silicon nitride to form a protective silica scale. In the gas turbine environment, the silica reacts with water vapor to form gaseous $Si(OH)_4$, which is swept away by the high velocity gas. The competing processes of scale formation and scale volatilization ultimately lead to a steady-state value of the scale thickness as well as a linear recession of the ceramic substrate. Using the approach outlined in [10], the steady-state thickness was predicted to be less than 1 μm for temperatures above 900°C. Scanning electron microscopy (SEM) confirmed that during the engine tests little or no silica formed.

The extensive scatter in the data in Figure 5 is most likely a consequence of the vane-to-vane temperature variations arising from the combustor pattern. To assess the influence of these variations, the recession was calculated as a function of time and temperature for a constant pressure (8.9 atm) and velocity (573 m/s) using the modified form of Equation 2,

$$k_l \text{ (}\mu\text{m/h)} = B \exp(-108 \text{ kJ/mole/RT)}v^{1/2}P_{total}{}^{3/2} \qquad (3)$$

where B is a constant. For the AS800 vanes, B was estimated by assuming that the maximum recession value measured at the midspan of the trailing edge after 815 h was associated with the vane subjected to the highest expected temperature of 2350°F (1288°C).[1] To verify the validity of this approach, the calibrated version of Equation 3 was then to used to estimate the temperatures for the remaining vanes. Table 1 summarizes the results. The calculated temperature rage is quite close to that expected on the basis of numerical analysis. The predicted time and temperature dependencies of the midspan, trailing edge recession (lines in Figure 5) are also in good agreement with the experimental data.

Dimensional measurements of Vane 24 made with the CMM revealed significant material loss in the region of the leading edge. The excessive recession in the nose region where the tangential velocity is low, appears to be in

---

[1] The recession rate is calculated by dividing the measured recession by the total time of 815 h.

contradiction with the velocity dependence exhibited by the model. To further investigate this effect, Vane 24 was sectioned as shown in Figure 4. The section representing the midspan was subsequently examined using SEM. Figure 6 compares the profile of the leading edge region in the as-received condition with that obtained after 815 h. The applicability of the recession model to predict the change in profile was examined by using Equation 1 to calculate the recession rate as function of position. The values of the pressure and velocity along the airfoil surface (required for this calculation) were obtained from the turbine manufacturer (Figure 5). These data were used in conjunction with Equation 3 to estimate the recession rate at each point along the airfoil surface. The change in the surface profile was calculated by assuming that the recession was normal to the surface at the point in question and that the magnitude was equal to the product of the rate and time. As shown in Figure 6, the prediction profile for the 815 h vane was in good agreement with the experimentally determined profile.

To further investigate the recession model, the loss of material along the trailing edge was measured for Vane 60. The temperature at each measurement point was determined from steady-state thermal analysis. Because of variations in combustion profiles, the temperature profiles varied with position. Figure 7 illustrates the case for the maximum temperature conditions. The trailing edge temperatures associated with the two extremes in temperature conditions were then used in conjunction with Equation 3 to predict the recession. Figure 8 compares the measured recession data with the two predicted profiles. Considering the uncertainty in predicted temperature values, the agreement is fairly good. More importantly, the experimental recession profile exhibits a sharp maximum near the midspan which is similar to the predicted curves.

The strengths of the ceramic vanes from Test 1 were measured using the ball-on-ring arrangement shown in Figure 9. Specimens were machined from both the airfoil and upper platform surface by first diamond core drilling small cylinders having nominal diameters of 5.5 mm. Each cylinder was then machined on one face only until the thickness was 0.4 to 0.5 mm. In this way one face of each specimen always consisted of the exposed surface of either the airfoil (convex side) or platform. During testing, this surface was loaded in tension.

The test fixture consisted of a 0.99 mm diameter hardened steel ball, which was mounted to a miniature load cell. The lower support ring, which was 5.0 mm in diameter, was fabricated from a high-strength polymer. It was mounted to a vertical stepper motor (Z stage) which was affixed to X-Y stages for positioning in the horizontal plane. All three stages were controlled by a computer. After placing a specimen on the lower support ring, the X-Y stages were used to position the assembly directly under the upper load ball. The Z-stage was then raised at a rate of 0.5 mm/s until the specimen made light contact with the ball. The specimen was subsequently loaded to failure at a displacement rate of 0.05 mm/s. The computer monitored and recorded the displacement, load, and time. The strength, S_b, was calculated from the equation

$$S_b = 3P(1+n)/(4nt^2) \bullet$$
$$[1 + 2ln(a/b) + ((1-n)/(1+n))(1 - b^2/2a^2)(a^2/R^2)] \quad (4)$$

where P is the ultimate sustained load, a is the radius of the support "ring", b is the effective radius of contact of the loading ball on the specimen, R is the specimen radius, t is the specimen thickness, and n is Poisson's ratio. As a first approximation, b was taken as t/3.

The strength data are summarized in Figure 10. The strengths of the specimens removed from the airfoil platform were consistently higher than those removed from the airfoil surface. This difference could in part be attributed to the better surface quality of the airfoil ends. There was very little change in average strength with time, although there was a small increase in the standard deviation in the strength of the specimens taken from the airfoil surface. These results indicate that the strength of the airfoil surface was not affected by the recession.

In Test 2, which is still ongoing, vanes included AS800 with an experimental environmental barrier coating (EBC) and uncoated SN282. Preliminary recession data (Figure 11) obtained from dimensional inspection of vanes periodically removed from the test indicate that the experimental EBC does not inhibit the environmental degradation.

SUMMARY

A key lesson learned in this work is that component properties are often quite different from those determined from standardized test specimens. The application of component characterization can address this limitation by providing mechanical data directly from the component in question. Furthermore, by evaluating properties as a function engine test time, important insights into the effect of environment upon performance can be obtained. Such insights are difficult is not impossible to generate in a standard laboratory environment.

Microstructural characterization of components has also been instrumental in establishing recession data. Existing models, which account for the competitive effects of silica volatilization and oxidation provide an excellent description of the pressure, velocity, and temperature dependence of the recession behavior. Environmental barrier coatings ultimately will be required to minimize this environmental effect.

REFERENCES

1. van Roode, M., Brentnall, W. D., Smith, K. O., Edwards, B. D., McClain, J., and Price, J. R., "Ceramic Stationary Gas Turbine Development - Fourth Annual Summary," ASME paper 97-GT-317, presented at the International Gas Turbine and Aeroengine Congress and Exposition, Orlando, Florida, June 2-5, 1997.
2. Wenglarz, R. A., Ali, S., and Layne, A.,"Ceramics for ATS Industrial Turbines," ASME Paper 96-GT-319, presented at the International Gas Turbine and Aeroengine Congress and Exposition, Birmingham, U.K., June 10-13, 1996.
3. Holowczak, J.E., Bornemisza, T.G., and Day, W.H., "Sector Testing of Cooled Silicon Nitride Ceramic HPT Vanes for the FT8 Aeroderivative Engine," in Proceedings of the ASME TurboExpo 2000, May 8-11, 2000 Munich, Germany, Paper Number 2000-GT-662.
4. Lin, H. T., Ferber, M. K., van Roode, M., "Evaluation of Mechanical Reliability of Si_3N_4 Nozzles after Exposure in an Industrial Gas Turbine, Design and Testing of Ceramic Components for Industrial Gas Turbines," pp. 97-102 in *Ceramic Materials and Components for Engines*. Edited by Heinrich JG, Aldinger A, Wiley-VCH, Weinheim-New York-Chicester-Brisbane-Singapore-Toronto, 2001.
5. Opila, E. J. and Hann, Jr., R. E., "Paralinear oxidation of SiC in water vapor," *J. Am. Ceram. Soc.*, **80** [1], 197-205, 1997.
6. Opila, E. J., "Variation of the oxidation rate of silicon carbide with vapor pressure," *J. Am. Ceram. Soc.*, **82** [3], 625-636, 1999.
7. Smialek, J. L., Robinson, R. C., Opila, E. J., Fox, D. S., and Jacobson, N., "Recession due to SiO_2 scale volatility under combustor conditions," *Advanced Comp. Mater.*, **8** [1]: 33-45, 1999.
8. Brentnall, W. D., van Roode, M., Norton, P. F., Gates, S., Price, J. R., Jimenez, O., and Miriyala, N., "Ceramic Gas Turbine Development at Solar Turbines Incorporated," to be published in Progress in Ceramic Gas Turbine Development, Volume I, ASME Press, eds. M. van Roode, D. Richerson, and M. K. Ferber.
9. Ferber, M. K., Lin, H. T., Parthasarathy, V., and Wenglarz, R. A., "Characterization of Silicon Nitride Vanes Following Exposure in an Industrial

Gas Turbine," presented at IGTI Meeting in Munich, Germany, May 2000; to be published as an ASME pamphlet.

10. Ferber, M. K., Lin, H. T., and Keiser, J., "Oxidation Behavior of Non-Oxide Ceramics in a High-Pressure, High-Temperature Steam Environment," *Mechanical, Thermal and Environmental Testing and Performance of Ceramic Composites and Components, ASTM STP 1392*, M. G. Jenkins, E. Lara-Curzio and S. T. Gonczy, Eds., American Society for Testing and Materials, West Conshohocken, PA, 2000.

Table 1. Summary of 815 h Recession Measurements and Estimated Temperatures.

Vane No.	Recession (μm) at Midspan of Trailing Edge After 815 h	T (°C)	T (°F)	Comments
3	76.2	983	1801	Calculated
5	127	1046	1915	Calculated
6	152.4	1070	1958	Calculated
8	203.2	1110	2031	Calculated
9	228.6	1127	2061	Calculated
10	254	1143	2089	Calculated
11	279.4	1158	2116	Calculated
12	304.8	1171	2140	Calculated
13	330.2	1184	2162	Calculated
14	355.6	1195	2184	Calculated
15	381	1207	2204	Calculated
16	406.4	1217	2223	Calculated
17	431.8	1228	2242	Calculated
18	457.2	1237	2259	Calculated
19	482.6	1246	2276	Calculated
24	609.6	1288	2350	Used to calibrate Eq. (1)

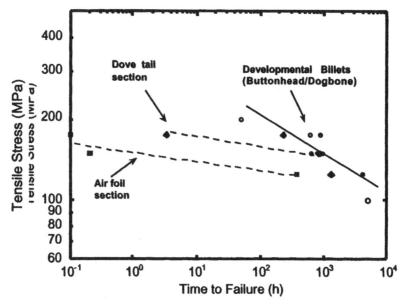

Figure 1: Comparison of stress rupture data generated for various sections of a ceramic blade with that obtained from standard specimens machined from developmental billets.

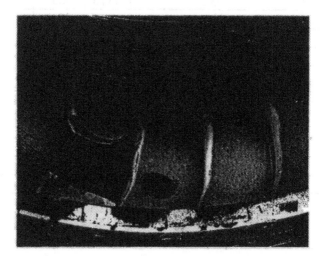

Figure 2: Nozzle assembly after 68 h showing failed nozzle.

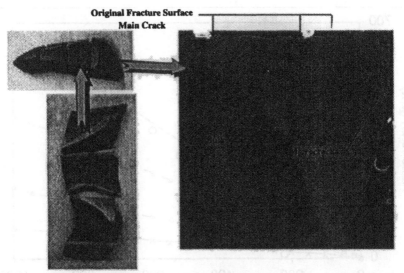

Figure 3: Cross-section of failed nozzle showing damage zone evolution in the vicinity of the primary fracture surface. The light-colored zone was found have a higher oxygen content compared to the bulk material.

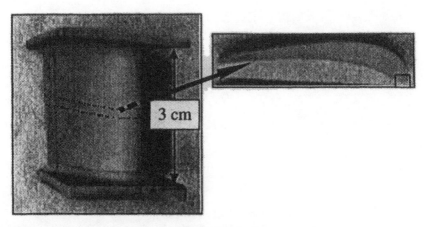

Figure 4: Sectioning of vanes required for SEM of cross-sections.

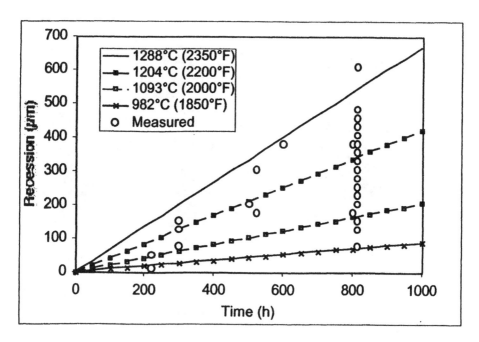

Figure 5: Measurement of recession of the silicon nitride vanes at the mid-span of the trailing edge (open circles). The lines in this figure represent predictions based upon a model described in the Introduction.

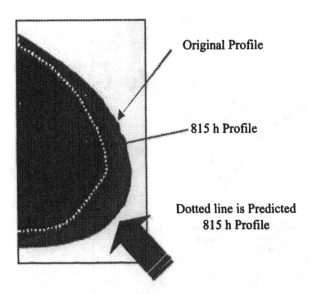

Figure 6: Profiles of leading edge (midspan) before testing and after 815 h. The large black arrow indicates gas flow direction.

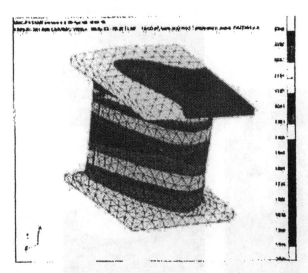

Figure 7. FEA results of steady-state temperature analysis of Rolls-Royce Allison Model 501-K turbine.

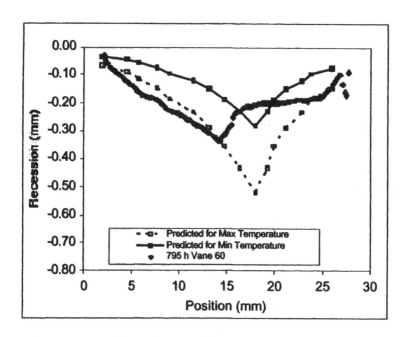

Figure 8. Comparison of measured and predicted recession along the trailing edge.

Figure 9: Biaxial test fixture and specimen.

Silicon-Based Structural Ceramics

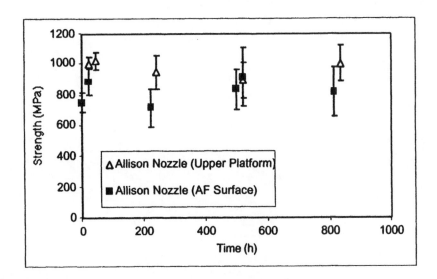

Figure 10 Biaxial flexure strength data versus engine exposure time.

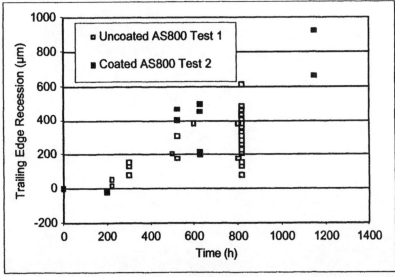

Figure 11: Recession in trailing edge (midspan) measured as a function time for both coated and uncoated silicon nitride vanes.

GELCASTING SIALON RADOMES

Mark A. Janney*, Donald M. Kupp**, Kevin W. Kirby , and Claudia A. Walls*
*Oak Ridge National Laboratory, Oak Ridge, TN.
**Fraunhofer Resource Center - Delaware, Newark, DE.
Hughes Research Laboratory, Malibu, CA.

INTRODUCTION

During the 1980's, General Dynamics - Pomona, CA developed an improved radome material[1] with properties superior to the standard radome material of the time, Corning's Pyroceram®. Known as GD-1 (i.e. General Dynamics-1), the material consisted of a β-SiAlON phase, Fig. 1, with the nominal composition represented by the formula $Si_{6-z}Al_zO_zN_{8-z}$, where $z \leq 2$. This material is isostructural with β-Si_3N_4, and tends towards some of its properties, Table 1.[2] The GD-1 composition was optimized for high temperature dielectric properties, especially low loss tangent at radar frequencies, while retaining reasonable strength, hardness, and thermal shock properties.

To produce GD-1 radomes, a non-aqueous slip-casting process was developed at General Dynamics. Non-aqueous processing was required to accommodate the AlN powder in the starting materials. While full-scale monolithic radomes were produced by this method, cracking and warping during the drying process consistently kept yields low. Consequently, despite the desirable material properties, the material processing difficulties forced General Dynamics to abandon radomes based on the GD-1 formulation.

In 1995, GD-1 fabrication was revisited in a collaborative effort between Hughes Missile Systems Company - Tuscon (the successor to General Dynamics - Pomona) and the Oak Ridge National Laboratory. Gelcasting was selected as the forming method.

Table 1. Comparison of radome material properties

Property	GD-1	Pyroceram	Silicon Nitride
Strength (25°C)	38 ksi [260 MPa]	30 ksi [210 MPa]	120 ksi [830 MPa]
Elastic Modulus	34×10^6 psi [230 GPa]	16×10^6 psi [110 GPa]	44×10^6 psi [300 GPa]
Knoop Hardness	1300	620	2200
Density	3.02 g/cm^3	2.6 g/cm^3	3.2 g/cm^3
Max use temperature	1300°C	1100°C	1500°C
Thermal expansion (1000°C)	4.1×10^{-6}	6.6×10^{-6}	4.3×10^{-6}
Thermal conductivity	16 W/m-K	4 W/m-K	22 W/m-K
Thermal shock (Figure of Merit, $\sigma\kappa/E\alpha$)	9940	2000	27000
Dielectric constant (1000°C)	7.0 - 7.7	5.5	7.96
Loss Tangent (1000°C)	0.003	0.03	0.013

GELCASTING PROCESS DEVELOPMENT

Gelcasting is a slurry-based ceramic forming process. A slurry comprising ceramic powder and a monomer solution is poured into a mold, polymerized *in-situ* to immobilize the particles in a gelled part, removed from the mold while still wet, then dried and fired. A flow chart for gelcasting is given in Figure 2. The details of the gelcasting process are given elsewhere.[3]

Table 2. Raw materials selection for GD-1

Composition	Slipcast GD-1 (1985)	Gelcast GD-1 (1995)
Si$_3$N$_4$	Cerac	Ube E-5
Al$_2$O$_3$	Cerac	RC-HP-DBM
AlN	Cerac	ART-WR100
Y$_2$O$_3$	Alfa-Aesar	Molycorp
SiO$_2$	Alfa-Aesar	Alfa-Aesar

Raw materials
The raw materials selected to develop the 1985 GD-1 formulation were based primarily on their high purity and not on their ceramic processing properties. This may have contributed to the difficulties encountered in slip casting these materials.

By 1995, ceramic powders had evolved significantly and one could select for both high purity and for processability. Hence, several of the raw materials were changed. Table 2 summarizes the original raw materials and those used for gelcasting.

The original GD-1 slip casting was done in a non-aqueous solvent to prevent hydrolysis of the AlN component of the slurry. By 1995, a water-processable AlN (AlN-WR100, ART Inc., Buffalo, NY) had become available and was selected for the gelcasting trials. Initial trials to determine the processing envelop for the water-resistant AlN showed that while AlN-WR100 could be processed in water, one still needed to be mindful that it was AlN and that it could react with water under certain conditions. First, prolonged mixing of AlN-WR100 in water had to be avoided. This was demonstrated when a jar mill exploded while being mixed over a weekend. The water-resistant coating provided protection to the AlN powder for 24h, since we had been mixing the slurry overnight. However, 72 hours was too long. The AlN reacted with the water in the slurry to form $Al(OH)_3$ and NH_3 gas, which built up extreme pressures and blew the cap off the polyethylene mixing jar. Thus, it was decided that the AlN fraction of the slurry would be the last component added and that it would be mixed for as short a time as possible after addition. Second, AlN-WR100 is resistant to attack by water at ambient temperature, but not at elevated temperature. Our original gelcasting trials with AlN-WR100 gelled the slurry at 50°C. Upon opening the mold, we noticed a strong ammonia smell and discovered that the slurry had turned to concrete. The AlN again had reacted with water to form a cementitious $Al(OH)_3$ phase, which had to be chipped out of the mold with a chisel. Thus, we decided that the GD-1 formulation would be gelled at ambient temperature.

Monomer selection - Because the radome was large (8.5 inch [21.6 cm] diameter at the base by 23 inches [58 cm] tall in the green state) and had a thin wall (0.25 inch [0.64 cm]), we needed to use a gel system that was as stiff and strong as possible. The monomer system MAM-NVP-MBAM (3:3:1, 20 wt % total monomer in solution) was chosen based on a previous study of gel strength and stiffness.[3]

Dispersants and defoamers- ART recommended Emphos CS1361(Witco Chemical, Houston) for AlN100WR; also required use of Foamaster VF (Witco Organics Div., Houston) as defoamer for foam caused by Emphos CS1361.

We used Dolapix PC33 (Zschimmer and Schwarz, Berlin, FRG) as the main dispersant for silicon nitride and alumina. The efficacy of Dolapix in silicon nitride suspensions had been demonstrated in an earlier study.[3]

Additional defoamers, Surfynol 104E (Air Products, Inc., Allentown, PA) and Dapro 1139 (Daniel Products, Co., Jersey City, NJ), were required to prevent and eliminate foaming during mixing and deairing. These were added both in the mill and to the deairing chamber.

<u>Initiator</u> - The initiator system ammonium persulfate (APS) with tetramethylethylene diamine (TEMED)was used to initiate free radical polymerization in this gelcasting system. We observed that adding the standard 10 wt% APS solution to the slurry caused localized coagulation in the slurry causing the formation of "clumps" in the slurry. The pH of the APS solution as-mixed is about 2. We hypothesized that the APS solution shifted the pH of the slurry to lower pH values and caused localized coagulation due to pH-induced flocculation. To mitigate this problem, we adjusted the pH of APS to pH 9 before adding it to the slurry. This effectively prevented coagulation.

<u>Large scale processing</u>
Each radome required about 4kg of ceramic to fill the mold, which translated into about 5 liters of suspension at 50 vol% solids. Mixing was accomplished in a 10-liter polyethylene mill using about 2 kg silicon nitride balls (1 cm diameter). All of the water, half of the dispersant, and half of the silicon nitride powder were added to the mill and mixed at high speed for four hours to disperse the silicon nitride. Additional silicon nitride was added to the mill along with additional dispersant in the ratios 1/4, 1/8, 1/16, and 1/16 until all of the silicon nitride had been added and dispersed. For each of the succeeding additions of silicon nitride, a period of slow rolling was required to wet out and disperse the powder. This was required because the slurry became dilatant after each aliquot of silicon nitride was added and slow rolling was the only way to incorporate the new powder into the existing slurry. Typically, the mill was slow rolled for 1/2 - 2 hours (2-5 RPM) to wet out the new powder. Then the speed of the mill was slowly increased to its maximum fast rolling speed (~100 RPM) over the course of 1-2 hours. The mill was kept at the fast speed for an additional 2 hours before the next addition of powder was made. Once all of the silicon nitride was well dispersed, the alumina, yttria, and silica were added. These materials were relatively easy to disperse. The last component added was the AlNWR100. The Emphos 1361 dispersant and the Foamaster VF defoamer were dissolved in the slurry then the AlN was added and mixed for about 2 h, again following the procedure of slow rolling followed by faster rolling.

Once the slurry was mixed, it was deaired. Additional defoamers were added at this point. Deairing was accomplished in a 40 liter container held within a vacuum chamber. The large deairing container was required because the slurry typically increased in volume by 5-6 times during deairing (up to 30 liters) before the foam broke.

After deairing, the slurry was chilled to 5°C. This was done to ensure sufficient working time for the large volume of slurry. Let us digress a moment. To ensure that the part could be gelled at ambient temperature, the initiator concentration was doubled relative to what was normally used for silicon nitride-based gelcasting systems. However, because such a large volume of slurry was required to cast the radome, significantly more working time than normal was also required, which was in conflict with the shorter working time induced by the higher initiator concentration. We estimated that we would need at least 20 minutes to mix the initiator and catalyst with the slurry, to deair the slurry after initiator addition to remove bubbles and dissolved oxygen, to screen the slurry after deairing to remove coagulated "clumps" and dried flakes, and to cast the slurry into the mold. We chose to resolve this conflict by chilling the suspension to about 5°C before adding the initiator. This gave us more than an hour of working time.

Casting
Casting the part was another adventure in gelcasting development. The mold for the radome weighed about 350 lb [160 kg] and therefore could not be easily moved. The mandrel weighed about 60 lb [27 kg] and was inserted after the slurry was poured into the mold cavity. We chose to fill the mold with a gravity feed system using an 8 liter plastic reservoir with a plastic tube that connected to the bottom of the reservoir. This allowed us to convey the slurry to the bottom of the mold which prevented entrapping air bubbles in the casting. After all of the slurry was transferred to the mold cavity, the mandrel was lowered into the slurry by means of a pulley system. The mandrel was secured in place using alignment pins and hold-down screws. The part was allowed to gel for about 2 hours at ambient temperature. Then the mandrel was withdrawn - first using jack screws to "break" the mandrel away from the part, then using the pulley system to fully withdraw the mandrel from the part.

Drying
Perhaps the biggest challenge of the project was drying the radome uniformly. The radome is in the shape of a thin-walled ogive or bullet-shaped cone. The as-gelled stiffness of the radome was far too low to hold its shape if it were to be taken out of the mold immediately after gelation. Therefore, it had to be partially dried in the mold. The challenge was to dry all regions of the radome at the same rate. Unfortunately, the nose of the radome invariably dried slower than the body. Because the tip was weak and pliable, it invariably cracked or broke off when the

radome was removed from the mold, Figure 3.

Several engineering approaches were attempted to solve these problems. First, we tried drying the radome very slowly over the course of four days by covering the open end with wet towels. The tip still dried much more slowly than the body. Second, an air lance was lowered into the radome and dry air was directed at the tip of the radome. Moisture was carried away as the forced air was swept out of the mold. Even with the open end of the radome covered with plastic to prevent evaporation by diffusion and natural convection, the tip still dried much slower than the body of the radome. Third, we imposed a temperature gradient on the mold using the heating channels included in the mold walls. Even with a temperature gradient of 20 degrees (30°C at the tip and 10°C at the open end) the tip still dried too slowly relative to the body of the radome to prevent cracking. All of these approaches were handicapped by the fact that the air in the narrow tip region of the radome rapidly became saturated with water vapor, which significantly reduced the rate of drying in that region.

The drying problem was finally solved by applying a radically different drying approach. Instead of conducting the initial, in-mold drying of the radome using air, we chose to use a liquid desiccant[4]. After the mandrel was removed from the mold, the mold was filled with polyethyleneglycol 400 (PEG400). The osmotic stress in the PEG400 was significantly higher than the osmotic stress in the gelcast gel; therefore, water was pulled out of the gelled part and into the PEG400. The PEG400, being a liquid, had access to all of the surfaces of the radome including the tip. Also, the chemical potential for water to move from the gel into the PEG400 was quite high and was not limited by saturation effects as had been the air drying approaches. After 1-1/2 hours, the radome had dried as much as it would under the influence of the liquid desiccant. The PEG400 was siphoned out of the mold. At this point, the radome was uniformly "leather hard" from base to tip and was strong and stiff enough to be safely removed from the mold. After removal, the radome was rinsed in cold water to remove excess PEG400, it was covered with towels, and allowed to air dry slowly over night . The next day the radome was uncovered, and it was dried further at ambient conditions (about 22°C and 50% RH) for another 24 hours. Final drying was accomplished at 50°C in a convection oven. It was essential to remove as much water as possible before the radome was finish dried at 50°C to prevent reaction of AlN with water. The final dried radome is shown in Figure 4.

Thermal Processing
To remove the polymer binder, the green body was fired in air to 500°C before going to higher temperatures for densification. This temperature was high enough

to remove all of the organics from the body, but no so high as to significantly oxidize the silicon nitride or aluminum nitride powders.

Final densification was accomplished at 1650°C in nitrogen at ambient pressure. Because GD-1 sinters by a liquid phase mechanism, it becomes quite plastic during sintering. This property was used to advantage to define the final shape of the radome. The parts were fired conformally on a graphite mandrel, which defined the inner surface of the radome and resulted in a close to net shape part. The dimensions and positioning of the mandrel are such that the thermal expansion of the graphite and shrinkage of the dome produce the desired contour and size upon completing the thermal treatment. Densification of the radome in the high temperature step typically involves a uniform dimensional change of ~ 19 %, such that the overall amount of shrinkage from the as-cast radome to the final product is only ~ 22 % (i.e. the as-cast dimension is 1.22 times larger than the final dimension). Once the high temperature firing has been completed, the radome is removed from the mandrel and examined for dimensional tolerance. While some machining has been necessary to obtain the required dimensions, the results have indicated that this step may be virtually eliminated with further development.

SUMMARY

Full size sialon radomes were fabricated using aqueous gelcasting. The keys to success included a water resistant AlN powder which enabled us to use an aqueous gelcasting system, mixing procedures which allowed us to achieve high slurry density, and liquid desiccant drying which made initial drying uniform especially in the tip region.

REFERENCES

1. R.F. Lowell, "Development of SIALON for Single Frequency and Broadband Radome Applications," *Proceedings of the 1st DoD Electromag Windows Symposium*, Vol. 1, pp 2-1-1 to 2-1-8 (1985).
2. K.W. Kirby, A. Jankiewicz, M.A. Janney, C.A. Walls, and D.M. Kupp, "Gelcasting of GD-1 Ceramic Radomes," *Proceedings of the 8th DoD Electromag Windows Symposium*, pp 287-99 (2000).
3. M.A. Janney, O.O. Omatete, C.A. Walls, S.D. Nunn, R.J. Ogle, and G. Westmoreland, "Development of Low-Toxicity Gelcasting Systems," *J. Am. Ceram. Soc.*, **81** [3] 581-91 (1998).
4. M.A. Janney and J.O. Kiggans, Jr., "Method of drying articles," U.S. Pat. No. 5,885,493, March 23, 1999.

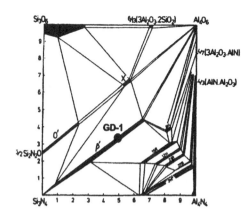

Figure 1. Quaternary phase diagram showing location of GD-1 composition.

Figure 3. The radome tip dried much slower than the body during air drying.

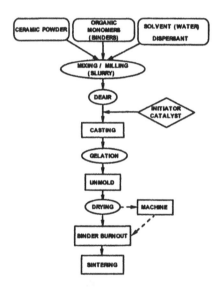

Figure 2. Gelcast processing is similar to other ceramic processes such as slip casting and spray drying.

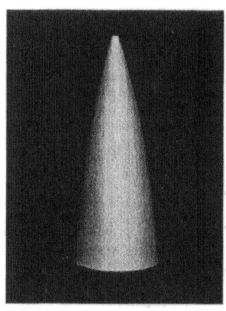

Figure 4. Green gelcast sialon radome. Base diameter = 8.5 in [22 cm], height = 23 in. [58 cm].

EFFECT OF LONG-TERM OIL IMMERSION TEST ON MECHANICAL RELIABILITY OF CANDIDATE SILICON NITRIDE CERAMICS FOR DIESEL ENGINE APPLICATIONS

H. T. Lin, T. P. Kirkland, and A. A. Wereszczak
Metals and Ceramics Division
Oak Ridge National Laboratory
Oak Ridge, TN 37831-6068

M. J. Andrew
Caterpillar Inc.
Technology Center
Peoria, IL 61656

ABSTRACT
Two lots of commercially available silicon nitride ceramics (i.e., Honeywell GS44 and Kyocera SN235) were exposed to an oil ash environment, as so-called oil immersion test, to evaluate the long-term corrosion/oxidation resistance in a simulated diesel engine environment. The exposure condition was at 850°C for 1000h in air. Subsequently, the exposed bend bars were tested for strength degradation at room temperature and 850°C at stressing rates of 30 MPa/s and 0.003 MPa/s in air, respectively. Little change in strength was measured after 1000h exposure to the mentioned oil ash environment. Also, the values of Weibull modulus obtained for all of the exposed silicon nitride materials were similar to those with as-machined surfaces tested under the same conditions. In addition, the detailed SEM/EDAX analyses indicated that no oil ash elements were detected in the bulk of materials, especially along the grain boundaries, resulting in no changes in chemistry of secondary phase(s). The oil ash elements were only present in a region about 1-3 μm below the exposed surface with no apparent changes in microstructure observed. Results of mechanical properties and microstructural characterizations indicated that these candidate silicon nitride materials exhibited excellent corrosion/oxidation resistance to the diesel engine environments, and would be excellent candidates for diesel engine exhaust valve applications.

INTRODUCTION

Advanced silicon nitride ceramics with reinforcing elongated grain microstructure continue to be of interest for use as exhaust valves in advanced diesel engines. Compared to conventional stainless steels used to fabricate current valves, silicon nitride is lighter, harder, and can withstand higher operating conditions. Implementation of silicon nitride valves in advanced diesel engines can potentially result in better efficiency, lower exhaust NOx and CO emission, and longer lifetime [1-4]. Rodgers, et al., [3] quantified this difference through the examination of a 2.8 liter overhead valve V-6 engine. A 20% increase in engine speed, a 30% reduction in the maximum valve forces, and a 30% reduction in valve train friction could be realized with a change to silicon nitride valves, and this would manifest itself into as much as a 5% increase in fuel economy.

An extensive testing program has been carried out to down select candidate silicon nitride ceramics for valve components and, also, for the generation of probabilistic strength and fatigue database on several commercially available silicon nitride ceramics [5]. Based on the test results and extensive microstructural analysis both Honeywell GS44 and Kyocera SN235 were then down selected for long-term corrosion test in a simulated diesel engine environment. The final down select of candidate silicon nitride material for exhaust valve manufacturing will be determined by the assessment of long-term stability of microstructure and chemistry under application environments. An oil immersion test was instrumented to evaluate the corrosion mechanisms of silicon nitride ceramics in heavy diesel engines [6]. This test will allow the testing of the effect of engine lubricants and their combustion products on the corrosion resistance of silicon nitride ceramics. It is anticipated that silicon nitride ceramics investigated would exhibit different corrosion mechanisms strongly depending upon the microstructure and chemistry of secondary phase(s).

The present study involved the generation of probabilistic strength and fatigue databases on two candidate silicon nitride ceramics after 1000h exposure to a simulated diesel engine environment (so-called oil immersion test). The characteristic strength, fatigue performance, and microstructural stability are compared and discussed.

EXPERIMENTAL PROCEDURES

The silicon nitrides tested in this study and used for comparisons are GS44 (Honeywell Ceramic Components, Torrance, CA) and SN235 (Kyocera Industrial Ceramics, Vancouver, WA). GS44 was manufactured via cold-isostatic-press and gas pressure sintering (GPS) with sintering additives of MgO, Y_2O_3, and Al_2O_3, while the Kyocera SN235 was fabricated via GPS with Y_2O_3, and Al_2O_3 sintering additives. Bend bar specimens with dimensions 3 mm x 4 mm x 50 mm of each material were machined and transversely ground from purchased billets per ASTM C1161 [7]. All specimens were longitudinally chamfered.

The test silicon nitride samples were place in platinum crucibles and covered with commercial 10W30 engine oil. The crucibles with oil-covered silicon nitride bend bars were heated in a furnace at 600°C for approximately 30 minutes to ash the oil. After completion of oil-ash conversion, the samples were then heat-treated at 850°C for 1000h in an ambient air. The ash could then act as a corrodent and attack the silicon nitride ceramics during long-term heat treatment. The so-called oil immersion test provides the most harsh corrosion/oxidation environment that the silicon nitride components will be subjected to during the engine operation.

Flexure testing was conducted in ambient air in four-point-bending using 20/40mm, □-SiC, semi-articulating fixtures at two conditions: 20°C and 30 MPa/s; and 850°C and 0.003 MPa/s. The first test condition was chosen to evaluate the effect of long-term oil immersion test on the inert strength, while the second test condition was chosen to measure the change in slow crack growth (SCG) susceptibility at 850°C after oil ash exposure. Pneumatic actuators were programmed to produce the desired loading (and corresponding stressing) rate via a personal computer. A resistance-heated furnace with molybdenum disilicide heating elements provided the 850°C temperature. Load was continuously measured as a function of time, and flexure strength was calculated using ASTM C1161 [7].

The accumulated strength data was then further analyzed. The strengths for each test set were fit to a two-parameter Weibull distribution using the program CERAMIC [8], which uses maximum likelihood estimation advocated in ASTM C1239 [9]. Reported results are uncensored because fractography analysis was not performed to identify strength-limiting flaws.

After long-term exposure in oil ash environment the surface morphology of both SN235 and GS44 were characterized using optical and scanning electron microscopy. X-ray analysis was also carried out to evaluate the phases formed arising from the corrosion/oxidation reactions. Following dynamic fatigue test, SEM analysis was carried out on fracture surfaces and polished cross-sections of selected bend bars to characterize the extent of corrosion/oxidation reaction as well as degradation mechanisms.

RESULTS AND DISCUSSION
Characterization of Exposed Materials

The surface morphologies of both GS44 and SN235 silicon nitride after 1000h oil ash exposure were evaluated using optical microscopy and SEM. SEM examinations showed that there was a significant amount of white ash on the bend bar surfaces that was loosely adhered. Also, remnants of light-colored scales were observed on the surfaces (Fig. 1a and 2a). Observations revealed that the original machining features, i.e., grinding ridges and roughness, were still apparent in most of the surface areas where no light colored scales were present (Fig. 1a and 2a). At higher magnification, there was noticeable surface oxidation, resulting from 1000 h exposure at 850°C in air (Fig. 1b and 2b). X-ray analysis of as-exposed bend bars indicated the presence of calcium zinc phosphate ($Ca_{19}Zn(PO_4)_{14}$), β-silicon nitride, and silica. The formation of calcium zinc phosphate (light colored scales) resulted from the oxidation of oil ash elements, i.e., Zn, Ca, P, Na, and S, which were commonly present in the form of additives [6]. The formation of silica was due to the oxidation of silicon nitride grains.

SEM examinations of polished cross section of as-exposed SN235 and GS44 silicon nitride revealed little changes in subsurface microstructure within 1-3 μm depth (Fig. 3). In addition, element mapping of both polished samples showed that the elements of Zn, Ca, P, Na, and S present in the engine oil were only detected in the subsurface region of 1-3 μm. Note that ingression of Ca and/or Na could significantly decrease the softening point and viscosity of glassy phase(s), thus possibly resulting in substantial degradation in mechanical performance at elevated temperatures. Therefore, results of little or no subsurface attacks by the oil ash elements suggested that the silicon nitride grains and secondary phases of SN235 and GS44 are very stable in oil ash environment.

Uncensored Weibull Strength Distributions

Results of characteristic strength and Weibull modulus for both GS44 and SN235 silicon nitride after 1000h oil immersion test are listed in Table 1 and Fig. 4 and 5. The results of as-machined bend bars tested under the same conditions are also listed for comparison (Table 1). Results of mechanical tests at room temperature indicated that both GS44 and SN235 silicon nitride exhibited little or no degradation in both inert strength and Weibull modulus after 1000h exposure to a simulated diesel engine environment. Also, both strength and Weibull modulus obtained at 850°C and 0.003 MPa/s for the exposed bend bars were similar to those obtained for the as-machined materials tested under the same condition. The observed similarities in high temperature mechanical propertied suggested that there was little change in secondary phase microstructure and chemistry arising from the ingression of oil elements during 1000 h exposure, consistent with SEM observations and element mapping results presented above.

Examinations of fracture surfaces of samples tested at 850°C indicated no change in surface feature of SN235 silicon nitride, while there was little formation of glassy phase and presence of pores in the 2-3 μm subsurface region in GS44 silicon nitride (Fig. 6), similar to the observations of polished cross section (Fig. 3). The softening point of secondary phase in GS44 is between 800 and 850°C, while the softening point of SN235 is between 1150 and 1200°C. It is expected that the long-term oil immersion test at 850°C in air could have more profound effect on mechanical performance of GS44 than SN235 due to its relative low softening point secondary phase. It was hypothesized that the oil ahs elements would readily ingress into the bulk of GS44 silicon nitride along the viscous glassy phase at grain boundaries, further lowering the viscosity and thus degradation of high temperature mechanical properties. However, SEM/EDAX

Silicon-Based Structural Ceramics

analysis indicated little changes in both microstructure and chemistry of as-exposed GS44 silicon nitride, thus no degradation in mechanical performance arising from oil ash exposure, as seen in Table 1. It is possible that the silica formation on surfaces due to the oxidation of silicon nitride grains could inhibit the ingression of oil ash elements into the bulk and secondary phase. Both the mechanical results and SEM examinations led one to down select the Kyocera SN235 as the final candidate for exhaust valve component applications. In addition, the depth of corrosion attack in metallic alloys ranges from 4 to 18 μm after 48h oil immersion test at 750°C [10], which further demonstrate that the silicon nitride ceramics exhibit superior corrosion resistance to current candidate metallic alloys. An engine test with SN235 silicon nitride ceramic valves would be carried out to validate the probabilistic design, mechanical performance, and corrosion resistance.

CONCLUSIONS

Flexural tests were carried out on both Kyocera SN235 and Honeywell GS44 bend bars after 1000h exposure to an oil ash environment at room temperature and 850°C at stressing rates of 30 MPa/s and 0.003 MPa/s in air. Little change in strength was measured after 1000h exposure to the mentioned oil ash environment. Also, the values of Weibull modulus obtained for all of the exposed silicon nitride materials were similar to those with as-machined surfaces tested under the same conditions. In addition, the detailed SEM/EDAX analyses indicated that no oil ash elements were detected in the bulk of materials, especially along the grain boundaries, resulting in no changes in chemistry of secondary phase(s). The oil ash elements were only present in a region about 1-3 μm below the exposed surface. Results indicate that these candidate silicon nitride materials exhibit excellent corrosion/oxidation resistance to the simulated diesel engine environment, which are far superior to those metallic alloys currently used for the diesel engine exhaust valves.

ACKNOWLEDGMENTS

Research sponsored by the U.S. DOE, Office of Transportation Technologies, Heavy Vehicle Propulsion System Materials Program, under Contract DE-AC05-00OR22725 with UT-Battelle, LLC.

REFERENCES

[1]   D. W. Richerson, <u>Modern Ceramic Engineering</u>, Marcel Dekker, Inc., New York, NY, 1982.
[2]   R. R. Wills, "Ceramic Engine Valves," *Communications of the American Ceramic Society*, 72 1261-1264 (1988).
[3]   G. Rodgers, R. Southam, J. Reinicke-Murmann, and P. Kreuter, "Analysis of Potential Improvements in Engine Behavior Due to Ceramic Valve Train Components," *Society of Automotive Engineers, Paper No. 900452* (1990).

[4]   R. Hamminger and J. Heinrich, J., "Development of Advanced Silicon Nitride Valves for Combustion Engines and Some Practical Experience on the Road," *Proceedings of the Materials Research Society Symposia*, Vol. 287, 513-518 (1993).

[5]   A. A. Wereszczak, H. -T. Lin, T. P. Kirkland, M. J. Andrews, and S. K. Lee "Strength and Dynamic Fatigue of Silicon Nitride at Intermediate Temperatures," *J. Mater. Sci.* 37 (2002) 2669-2684.

[6]   H. M. Abi-Akar, A Test Method to Evaluate High Temperature Corrosion," Paper No. 00530, Corrosion 2000, NACE International Annual Meeting, Houston, TX (2000).

[7]   "Standard Test Method for Flexural Strength of Advanced Ceramics at Ambient Temperatures," ASTM C1161, Annual Book of ASTM Standards, Vol. 15.01, American Society for Testing and Materials, West Conshohocken, PA, 1999.

[8]   "Life Prediction Methodology for Ceramic Components of Advanced Heat Engines, Phase 1," Prepared by AlliedSignal Engines, Phoenix, AZ, ORNL/Sub/89-SC674/1-2, DOE Office of Transportation Technologies, 1995.

[9]   "Practice for Reporting Uniaxial Strength Data and Estimating Weibull Distribution Parameters for Advanced Ceramics," ASTM C1239, *Annual Book of ASTM Standards, Vol. 15.01*, American Society for Testing and Materials, West Conshohocken, PA, 1999.

[10] M. Andrews, Caterpillar Inc., private communication.

Table I. Summary of uncensored Weibull strength distributions for GS44 and SN235 silicon nitride specimens, transversely machined per ASTM C1161, in as-machined and after 1000 h oil immersion test condition.

Material	# of Spmns. Tested	Stressing Rate (MPa/s)	Temp. (°C)	Uncens. Weibull Modulus	± 95% Uncens. Weibull Modulus	Uncens. Chrctstic Strength (MPa)	± 95% Uncens. Chrctstic Strength (MPa)
GS44	15	30	20	28.9	17.5, 43.0	793	775, 811
GS44	15	0.003	850	25.8	16.6, 37.0	545	532, 556
GS44*	10	30	20	39.20	21.8, 63.5	786	771, 800
GS44*	10	0.003	850	23.04	13.3, 35.8	488	472, 503
SN253	15	30	20	23.8	15.4, 33.9	901	879, 923
SN235	14	0.003	850	18.5	11.8, 26.8	744	720, 767
SN235*	10	30	20	18.64	10.24, 29.97	894	858, 929
SN235*	10	0.003	850	13.04	6.92, 21.52	827	777, 877

*specimens exposed to an oil ash environment at 850°C for 1000 h in air

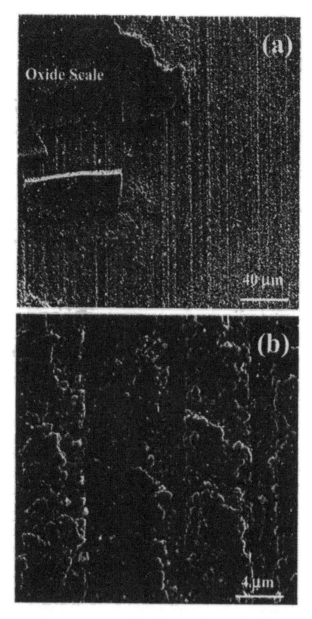

Fig 1. SEM surface features of Kyocera SN235 after 1000h oil immersion test. (a) Photo shows the light-colored scale (calcium zinc phosphate) and machining marks. (b)Photo reveals apparent oxidation and silica formation.

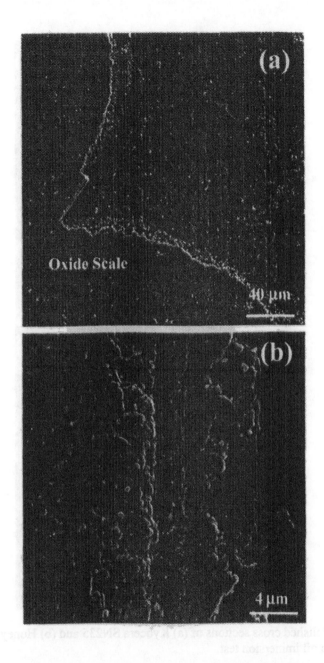

Fig 2. SEM surface features of Honeywell GS44 after 1000h oil immersion test. (a) Photo shows the light-colored scale (calcium zinc phosphate) and machining marks. (b)Photo reveals apparent oxidation and silica formation.

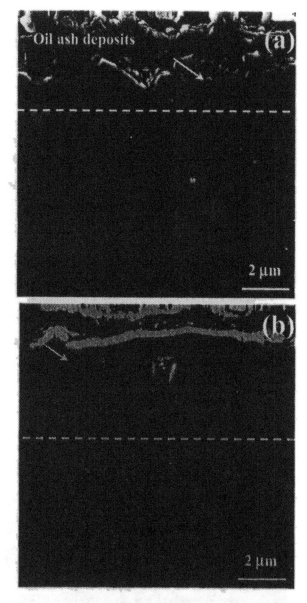

Figure 3. Polished cross sections of (a) Kyocera SN235 and (b) Honeywell GS44 after 1000h oil immersion test.

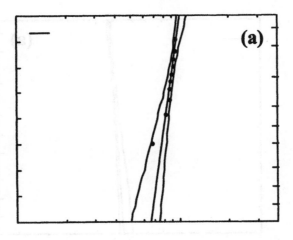

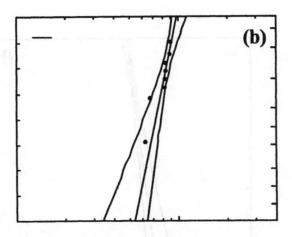

Figure 4. Uncensored flexural strength distribution of exposed Kyocera SN235 tested at (a) 20°C and 30 MPa/s and (b) 850°C and 0.003 MPa/s.

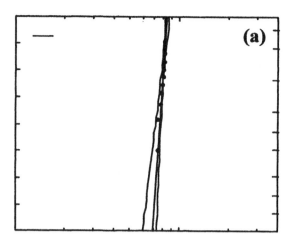

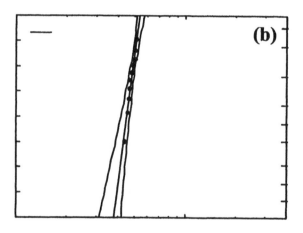

Figure 5. Uncensored flexural strength distribution of exposed Honeywell GS44 tested at (a) 20°C and 30 MPa/s and (b) 850°C and 0.003 MPa/s.

Silicon-Based Structural Ceramics

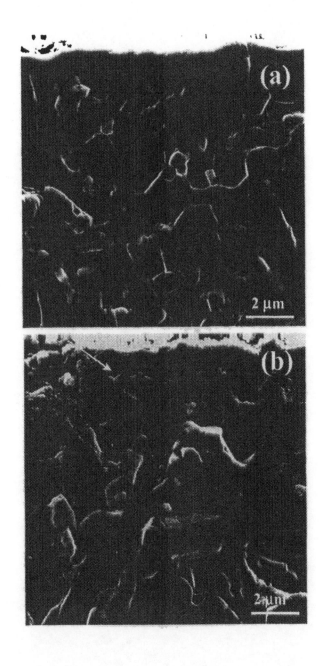

Figure 6. Fracture surfaces of exposed (a) Kyocera SN235 and (b) Honeywell GS44 samples after flexural testing at 850°C and 0.003 MPa/s.

KEYWORD AND AUTHOR INDEX

Michael, G., 213
Microchannels, 225
Microstructural development, 91, 203
Microstructure, 161, 177
Miyamoto, Y., 63

Nagel, A., 161
Neodymia, 161
Nitriding, 63

Ohji, T., 177
Ohta, T., 77
Oil immersion test, 261
Omori, M., 77
Ota, K., 113
Oxidation resistance, 261

Pabit, E.L., 225
Pezzotti, G., 113
Porous ceramic, 77, 177
Powder, SiAlON, 77
Powders, silicon nitride, 3, 47
Pu, X.-P., 35

Quist, M., 225

Radomes, 253
Radwan, M., 63
Rare-earth oxides, additives, 191, 203
Ray, D., 225
Rheology, 35
Roebben, G., 123
Rundgren, K., 47

Šajgalík, P., 191
Sand, 63
Schilm, J., 213
Shimada, S., 63
SiAlON, 77, 161, 253
Silicon carbide, 3, 91, 123, 191, 203, 225

Silicon nitride, 3, 35, 47, 91, 123, 145, 177, 213, 237, 261
Silicon oxynitride, 63
Sinterability, 161
Sintering, liquid phase, 91, 191, 203
Sintering, spark plasma, 77
Spark plasma sintering, 77
Spectroscopy, mechanical, 113, 123
Stiffness, 123
Swift, G.A., 145

Tatami, J., 77
Turbine, gas, 237

Üstündag, E., 145

Van Der Biest, O., 123
Viscoelastic properties, 35

Walls, C.A., 253
Wereszczak, A.A., 261
Wilson, M.A., 225

Yamauchi, Y., 203
Yang, J.-F., 177
Ytterbia, 161

Zhang, C., 77
Zhou, Y., 203

Printed and bound by CPI Group (UK) Ltd, Croydon, CR0 4YY

16/04/2025

14658442-0003